ZO121

ANIMAL DIVERSITY - II

[2 Credits]

For

F.Y.B.Sc. Zoology, Paper - I, Semester II
As Per New Revised Syllabus, CBCS Pattern
From June 2019

PRIN. DR. KISHORE R. PAWAR
M.Sc., Ph.D.
Former Principal,
Karmveer Shantarambapu Kondaji Wavare,
Arts, Sci. & Com. College, CIDCO,
NASHIK – 422008.

DR. ASHOK E. DESAI
M.Sc., Ph.D.
Former Associate Professor &
Head, P.G. Deptt. of Zoology,
K.T.H.M. College,
NASHIK – 422002.

N5034

F.Y.B.Sc. ANIMAL DIVERSITY - II   ISBN 978-93-89533-66-8

First Edition : November 2019
© : Authors

Published By:
NIRALI PRAKASHAN
Abhyudaya Pragati, 1312, Shivaji Nagar
Off J.M. Road, PUNE – 411005
Tel - (020) 25512336/37/39, Fax - (020) 25511379
Email : niralipune@pragationline.com

➢ **DISTRIBUTION CENTRES**

PUNE

Nirali Prakashan : 119, Budhwar Peth, Jogeshwari Mandir Lane, Pune 411002,
(For orders within Pune) Maharashtra, Tel : (020) 2445 2044, Mobile : 9657703145
Email : niralilocal@pragationline.com

Nirali Prakashan : S. No. 28/27, Dhayari, Near Asian College Pune 411041
(For orders outside Pune) Tel : (020) 24690204; Mobile : 9657703143
Email : bookorder@pragationline.com

MUMBAI

Nirali Prakashan : 385, S.V.P. Road, Rasdhara Co-op. Hsg. Society Ltd.,
Girgaum, Mumbai 400004, Maharashtra;
Mobile : 9320129587 Tel : (022) 2385 6339 / 2386 9976,
Fax : (022) 2386 9976
Email : niralimumbai@pragationline.com

➢ **DISTRIBUTION BRANCHES**

JALGAON

Nirali Prakashan : 34, V. V. Golani Market, Navi Peth, Jalgaon 425001,
Maharashtra, Tel : (0257) 222 0395, Mob : 94234 91860;
Email : niralijalgaon@pragationline.com

KOLHAPUR

Nirali Prakashan : New Mahadvar Road, Kedar Plaza, 1st Floor Opp. IDBI Bank,
Kolhapur 416 012, Maharashtra. Mob : 9850046155;
Email : niralikolhapur@pragationline.com

NAGPUR

Nirali Prakashan : Above Maratha Mandir, Shop No. 3, First Floor,
Rani Jhanshi Square, Sitabuldi, Nagpur 440012, Maharashtra
Tel : (0712) 254 7129;
Email : niralinagpur@pragationline.com

DELHI

Nirali Prakashan : 4593/15, Basement, Agarwal Lane, Ansari Road, Daryaganj
Near Times of India Building, New Delhi 110002
Mob : 08505972553, Email : niralidelhi@pragationline.com

BENGALURU

Nirali Prakashan : Maitri Ground Floor, Jaya Apartments, No. 99, 6th Cross,
6th Main, Malleswaram, Bengaluru 560003, Karnataka;
Mob : 9449043034
Email: niralibangalore@pragationline.com

Other Branches : Hyderabad, Chennai

niralipune@pragationline.com | www.pragationline.com

Also find us on www.facebook.com/niralibooks

Preface ...

It gives us great pleasure to present this book "Animal Diversity – II for the students of First Year B.Sc. Zoology Second Semester. This book is written according to the new syllabus as per CBCS Pattern of June 2019.

We have tried our best to present the subject matter in an easy style and in a comprehensive manner. The subject matter is profusely illustrated with a number of clear and labelled diagrams. We sincerely feel that this book will fulfill the requirements of the students as well as teachers. While preparing this book several standard reference books and text books have been consulted. Emphasis has been laid on furnishing maximum information required for students in a simple and lucid language. Zoology is an interesting subject because the animal world is full of diversity, adaptations, habits and habitats and behaviour. There are several textbooks and reference books available written by Indian and foreign authors but these books are too costly and majority of the students are unable to purchase them for comprehensive study.

We express our sincere thanks to Shri. Dineshbhai Furia, Shri. Jignesh Furia, Mr. Malik Shaikh, Mrs. Anjali Muley, Mrs. Roshan Shaikh and the entire staff of Nirali Prakashan for taking keen interest in the publication of this book and bringing out the book on time.

We shall gratefully accept constructive suggestions from the teachers as well as students for improvement of this book.

November, 2019   **Prin. Dr. Kishore R. Pawar**

Dr. Ashok E. Desai

Syllabus ...

1. Phylum Aschelminthes (04)

1.1 Introduction to phylum Aschelminthes

1.2 Salient Features of Phylum Aschelminthes

1.3 Classification of Phylum Aschelminthes (Class Nematoda only with two examples – *Ascaris lumbricoides* (common round worm), *Wuchereria bancrofti* (Elephantiasis)).

1.4 Economic importance of class Nematoda.

2. Phylum Annelida (06)

2.1 Introduction to Phylum Annelida

2.2 Salient Features of Phylum Annelida.

2.3 Classification of Phylum Annelida up to classes with examples of following classes (names of examples only).

Class Polychaeta (e.g: *Nereis pelagica*, Neries/sand worm)

Aphrodita aculeata (Aphrodite/sea mouse)

Class Oligochaeta (e.g. *Pheritima posthuma* (earthworm),

Class Hirudinea (e.g. *Hirudinaria granulosa*, common cattle leech)

2.4 Economic importance of Annelida with reference to earthworms as friends of farmers and their role in vermicomposting.

3. Phylum Arthropoda (06)

3.1 Introduction to Phylum Arthropoda

3.2 Salient Features of Phylum Arthropoda

3.3 Classification of Phylum Arthropoda with specific classes and mentioned examples (names only)

Class : Crustacea : *Palaemon palaemon* (Prawn) *Brachyura* spp. crabs)

Class : Chilopoda : *Scolopendra* sp. (centipede)

Class : Diplopoda : *Julus* sp. (millipede)

Class Insecta : *Periplaneta americana* (American Cockroach),

Anopheles stephensii (mosquito).

Class : Arachnida - Spiders, *Buthus sp.* (scorpion)

3.4 Mouth Parts in Insects: Mandibulate (Cockroach), Piercing and Sucking (Female Anopheles Mosquito), Chewing and Lapping Type (Honey Bee)

3.5 Economic Importance of Arthropoda

Useful Insects : Honey bee, Lac insect, Silkworm.

Harmful insects : Female Anopheles mosquito, Red cotton bug, Rice weevil

4. Phylum Mollusca (06)

4.1 Introduction to Phylum Mollusca

4.2 Salient features of Phylum Mollusca

4.3 Classification of Phylum Mollusca with specific classes and mentioned examples (names only)

Class Gastropoda e.g. *Pila globosa* (apple snail)

Class Pelecypoda e.g. *Lamellidens marginalis* (Bivalve)

Class : Polyplacophora e.g. *Chiton*

Class : Cephalopoda e.g. *Octopus vulgaris* (common octopus), *Sepia officinalis* (common Cuttle fish)

4.4 Economic Importance of Mollusca

5. Study of Phylum Echinodermata (08)

5.1 Introduction to Phylum Echinodermata

5.2 Salient Features of Phylum Echinodermata.

5.3 Classification of Phylum Echinodermata with specific classes and mentioned examples (names only)

Class : Asteroidea (*Asterias rubens,* sea stars or starfish)

Class : Holothuroidea. *Holothuria* sp. sea cucumbers)

Class : Echinoidea (*Echinus esculentis common* sea urchins)

Class : Crinoidea (sea lilies or feather stars)

5.4 Type Study: *Asterias rubens* (Sea Star): Classification, Habit, Habitat, External Morphology, Digestive System, Water Vascular System and autotomy and regeneration

5.5 Pedicillaria in Echinodermata: straight, crossed, valvate, tridactylous, globigerous.

5.6 Economic Importance of Echinidermata.

***

Contents ...

****

PATTERN OF UNIVERSITY EXAM QUESTION PAPER

Total Marks : 35   **Duration : 2 Hours**

Note :
  (a) Q. 1 is compulsory.
  (b) Attempt any three questions from Q.2 to Q.5.
  (c) Questions 2 to 5 carry equal marks.

1. **Attempt any five of the following :**

  (a)

  (b)

  (c)

  (d)

  (e)

  (f)

  (Ask four tricky questions and two questions based on problem solving, if applicable.)   (5)

2. (A) Describe – type of question(s) with internal options.   (6)

  (B) Short question, but tricky.   (4)

3. (A) Explain – type of question(s) with internal options.   (6)

  (B) Problem based question, if applicable otherwise justification type – question.   (4)

4. (A) Discuss - type of question(s) with internal options.   (6)

  (B) Problem based question, if applicable otherwise tricky and signified question.   (4)

5. **Write short notes on any four of the following :**   (10)

  (A) Principle based.

  (B) System based.

  (C) Structure based.

  (D) Descriptive based.

  (E) Working based.

  (F) Model based.

✱✱✱

1

CHAPTER

PHYLUM - ASCHELMINTHES

CONTENTS

1.1 INTRODUCTION TO PHYLUM ASCHELMINTHES

There are different opinions among zoologists about the systematic position of Nematodes. Genebaur and others have considered it as a separate Phylum Nematohelminthes which includes or does not include Acanthocephala and Nematomorpha. Zoologist Grobben coined the term Aschelminthes.

The Aschelminthes *(Greek askes = cavity; helmins = worm)* are pseudocoelomote, mostly vermiform fresh water and marine animals. Recently, Nematodes have been grouped in a class Nematoda in the phylum Aschelminthes.

1.2 FEATURES OF PHYLUM ASCHELMINTHES

1. Body of round worms is elongated, cylindrical, vermiform with tapering ends.
2. Body is unsegmented or superficially segmented covered with tough cuticle.
3. They are mostly minute to small size but some are of great length.
4. Caudal end of the body is generally straight in female but coiled in males and males are shorter than females.
5. Mouth is terminal surrounded by lips. In strongyloides the lips are modified into teeth known as leaf crown.
6. The sensory organs are amphids and papillae which are of great taxonomic value in ease of free-living forms.
7. Body is covered by rough resistant cuticle with bristles, spines, warts and papillae, etc.
8. Body cavity is pseudocoel.

(1.1)

9. Digestive tract is well developed generally made up of mouth, buccal cavity, pharynx or oesophagus, intestine and anus.
10. Nervous system consists of a nerve-ring encircling the oesophagus from it, nerves are given out anteriorily and posteriorly.
11. Protonephridia are absent but excretory system is made up of canals.
12. Sexes are separate.
13. They are ovo-viviparous, oviparous or viviparous.
14. Life cycle is complicated with or without intermediate host.
15. The phylum includes free-living, epizoic and parasitic members.

1.3 CLASSIFICATION OF PHYLUM ASCHELMINTHES

Phylum Aschelminthes is divisible into one Class - Nematoda.

Class Nematoda:

Distinctive Characters

1. Body is cylindrical, tapering towards both the ends.
2. Body is covered with cuticle.
3. Intestine is well developed.
4. Body cavity is not lined with epithelium.
5. Cloaca is absent in female.
6. Male and female reproductive organs are well developed.

    Examples: *Rhabdities, Loa loa, Dracunculus medinesis, Trichinella spirallis, Ascaris lumbricoides, Ancylostoma, Wuchereria.*

1.4 ECONOMIC IMPORTANCE OF ASCHELMINTHES

Nematoda: Several nematode worms are the parasites of man. There are four major prominant nematodes namely, round worms, hook worms, pin worms and whip worms that affect human. Ascariasis is disease caused by the most common worm *Ascaris lumbricoides*. Children are common victims. They are acquired in human body through contaminated food, water, raw vegetables, fruits. The adults reside in the intestine and patient complains of abdominal pains, weakness, vomitting, headache, dizziness, nervous disorder, skin rashes and fever often patient grits his teeth in sleep. They feed on the digested food, proteins of the victim and cause malnutrition among children. *Ancylostoma duodena* is called hookworm, causing disease Ancylostomiasis. They are common parasites within the intestine. They are common in rural areas. Children suffer more from these parasites. They are swallowed with contaminated food. These parasites suck blood and tissue juices of host. They enter blood vessels and are carried to the heart and lungs. The symptoms seen in the patient are gastrointestinal disorders, anaemia, nervous disorders.

Enterobiosis is caused by pinworm called *Enterobius*. They lay the eggs near the anal ring where they cause intense itching. They are picked upon the fingers and under the nails where they find their way to food and are swallowed. They hatch in stomach and juveniles migrate to the colon and develop into adult worms. Symptoms shown by the patient are loss of appetite, sleeplessness, bed wetting, grinding of teeth, nausea and vomiting.

Trichinella spiralis, the trichina worm cause disease called Trichinosis. It is transmitted by raw meat, especially pork. The symptoms of this disease are nausea, vomiting, odema of face and eyelids and fever.

Threadworms cause Strongyloidosis disease. The Elephantiasis or Filariasis is disease caused by filaria worm *Wuchereia bancrofti*. They live in the lymphatic system and connective tissues of the body and through which they circulate in the blood at night. Infection is spread through *Culex* mosquito. The worm causes enlargement of the limbs, scrotum and mammae. Swelling takes place due to blockage of lymph circulation by parasitic worms.

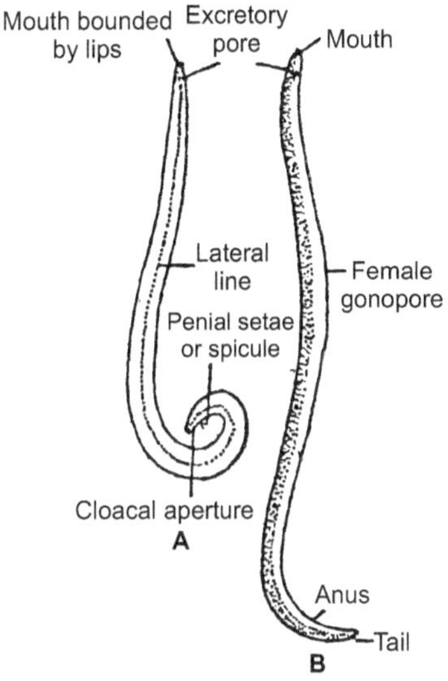

Fig. 1.1: *Ascaris lumbricoides*

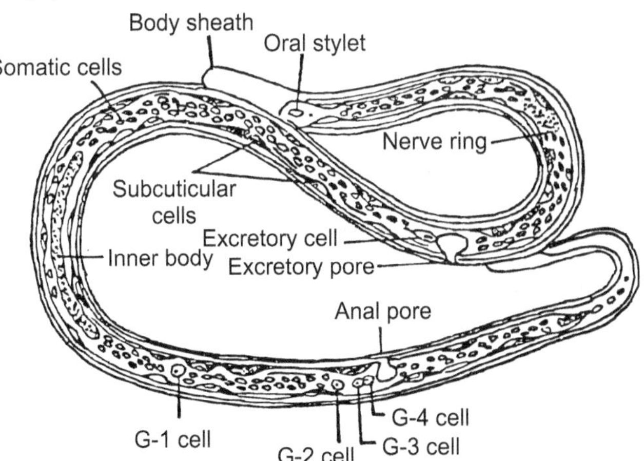

Fig. 1.2: *Wuchereria*. Microfilaria

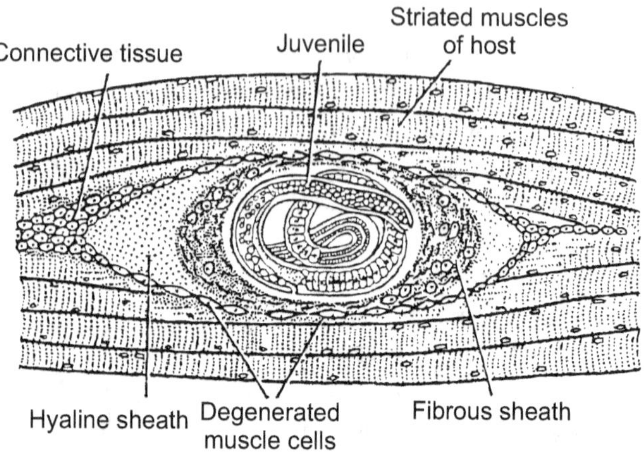

Fig. 1.3 : *T. Spiralis* Ensheated in Host Muscle

POINTS TO REMEMBER

- Aschelminthes are pseudocoelomate, vermiform freshwater and marine animals.
- Nematoda is only one class in this phylum.
- Body is covered by tough cuticle.
- Males are shorter than females.
- Mouth is terminal and surrounded by lips.
- Digestive tract is well developed.
- Excretory system is made up of canals.
- Life cycle is complicated with or without intermediate host.
- Several nematodes are human parasites.
- The parasites are roundworms, hookworms, pinworms and whipworms.
- *Ascaris* is common intestinal parasite of man.
- *Trichina spiralis* causes trichosis.
- *Wuchereria bancrofti* causes filariasis in man.

EXERCISE

1. Give an account of classification and salient features of phylum Aschelminthes.
2. Give an account of economic importance of Class Nematoda.
3. Write short notes on :
   (a) Nematoda.
   (b) Economic importance of Nematoda.
   (c) Important characters of Nematoda

2

CHAPTER

PHYLUM - ANNELIDA

CONTENTS

2.1 INTRODUCTION TO PHYLUM ANNELIDA

Lamark first used the name annelida for higher segmented worm means 'little ring' and refers to the ring like constrictions of the body. The name annelida has been derived either from Latin annelus (a ring), or French anneler (to arrange in rings) and the Greek eidos (form).

2.2 SALIENT FEATURES OF PHYLUM ANNELIDA

(1)   Most of the annelids are aquatic, fresh water as well as marine, some are terrestrial which live in burrows or in tubes.

(2)   Body is elongated, vermiform and bilaterally symmetrical.

(3)   Body is metamerically segmented. Externally, segments are shown by transverse grooves and internally by muscular partitions called septa. The segments or metameres or somites are many and arranged one after the other in a single linear series.

(4)   The outer most covering of the body is called cuticle secreted by underlying epidermal cells.

(5)   The body wall is composed of circular and longitudinal muscles.

(6)   Setae or chaetae are the locomotary organs.

(7)   They are triploblastic animals.

(8)   All annelids are true coelomate animals. The body cavity or coelomic cavity is lying between the two layers of mesoderm.

(9)   Respiration is by general body surface or by special projections or gills of parapodia and head.

(2.1)

(10) Annelids shows well developed blood vascular system and it is of a closed type.

(11) Blood is red due to dissolved haemoglobin.

(12) Nephredia are the excretory organs which communicate the coelom with the exterior end.

(13) Nervous system consists of cerebral ganglion (brain), circumpharyngeal connectives and double ventral nerve cord with segmental ganglia.

(14) Sexes may be separate (Polychaeta) or united (hermaphrodite) e.g. Oligochaeta and Hirudinea.

(15) In indirect development, there is characteristic larva called trochophore larva.

2.3 CLASSIFICATION OF PHYLUM ANNELIDA

The phylum Annelida has three main classes, such as,

(i) Polychaeta, (ii) Aelosomata, (iii) Clitellata

Class - Polychaeta:

Distinctive Characters of Polychaeta:

(1) Most of the animals are marine, few are fresh water and carnivorous.

(2) Body is elongated, cylindrical and segmented into similar somites.

(3) The animals show distinct head with sense organs such as eyes, tentacles, cirri, palps and mouth.

(4) Each body segment bears paired, flattened lateral outgrowths of body wall, called parapodia. They are locomotory and respiratory in function.

(5) Many bristles or setae extend from parapodia in bundles, hence animals are called polychaeta.

(6) The clitellum or cingulum is absent.

(7) Cirri or branchiae may be present on body segments.

(8) Sexes are generally separate (dioecious).

(9) Many shows asexual reproduction like serial or lateral budding.

(10) Fertilization is external.

(11) Development shows metamorphosis with a typical free swimming trochophore larva.

Examples: *Nereis, Aphrodite, Chaetopterus, Terebella, Amphitrite.*

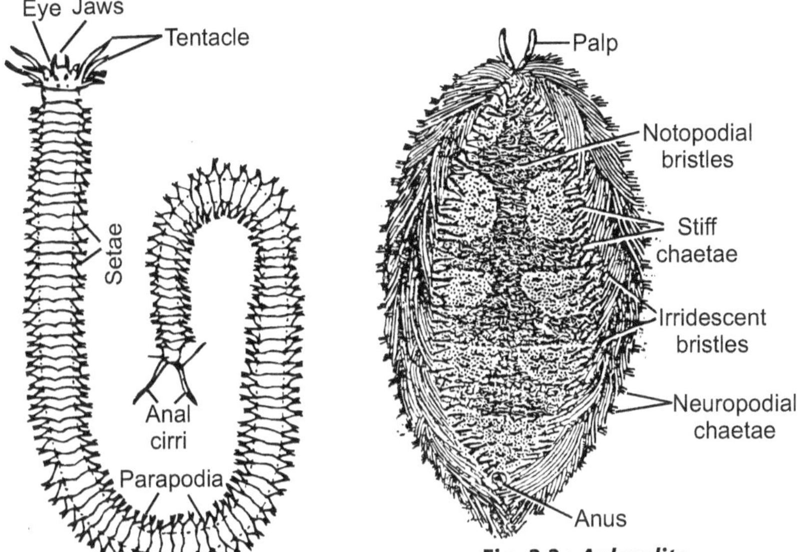

Fig. 2.1: *Nereis*. Dorsal View

Fig. 2.2 : *Aphrodite*

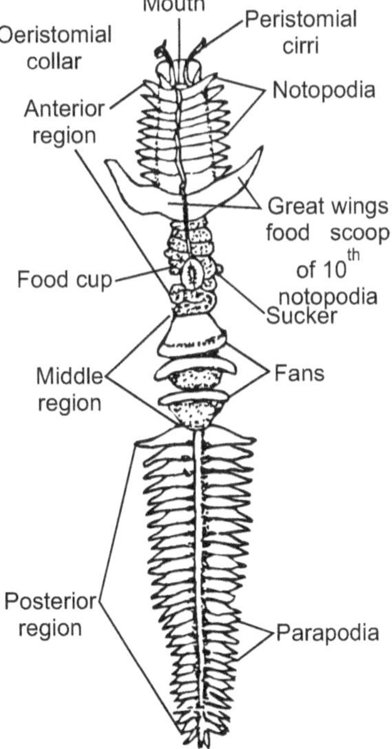

Fig. 2.3: *Chaetopterus*

Class-Aelosomata:

(1) These are small to minute worms with many chaetae.
(2) They live in the interstitial zone of both freshwater and brackish water environment.
(3) They are hermaphrodites animals.
(4) Each animal possessing one ovary and two testis.
(5) There are about 25 species. They are little known to science and their classification is disputed with some authors considering them to be part of the Oligochaeta.

Class-Clitellata is divided into three subclasses like **Oligochaeta**, **Hirudinaria** and **Brachiobdella**.

Sub-class-Oligochaeta:

Distinctive Characters of Oligochaeta:

(1) Most of the animals are terrestrial, some are fresh water and marine.
(2) No distinct head, prostomium is small without sense organs and appendages.
(3) Rod like setae are present in the body wall of each segment for locomotion.

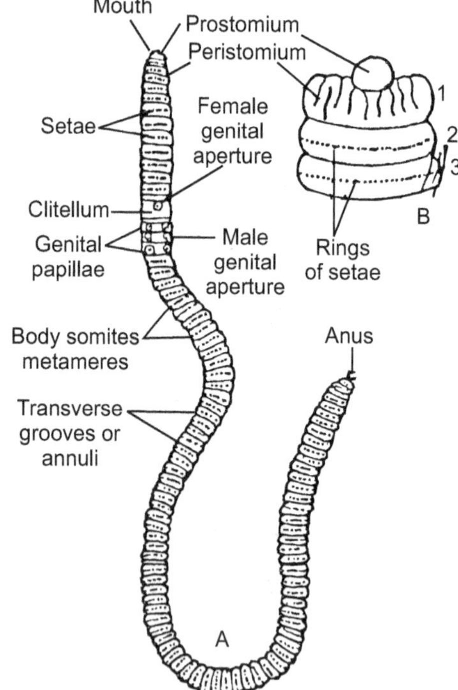

Fig. 2.4: *P. posthuma*. A – Entire worm in ventral view.
B – Anterior end in dorsal view

(4) Clitellum or cingulum is usually present.

(5)   Some animals bears external gills for respiration. Integument or skin also shows respiration.

(6)   They are bisexual or monoecious or hermaphrodite.

(7)   Development is direct. No metamorphosis and larval stage. Development occurs inside cocoon secreted by clitellum.

(8)   Some animals show asexual reproduction by transverse fission. All possess a great power of regeneration.

Examples: *Pheretima, Tubifex, Lumbricus, Branchiobdella.*

Sub-class-Hirudinea:

Distinctive Characters of Hirudinea:

(1)   Some are terrestrial, some freshwater and a few marine.

(2)   They are generally ectoparasitic and blood sucking.

(3)   Body is dorso-ventrally flattened or cylindrical, segmented.

(4)   Number of body segments is small but definite usually 33. Each segment again shows superficial divisions into 2 to 5 transverse rings or annuli.

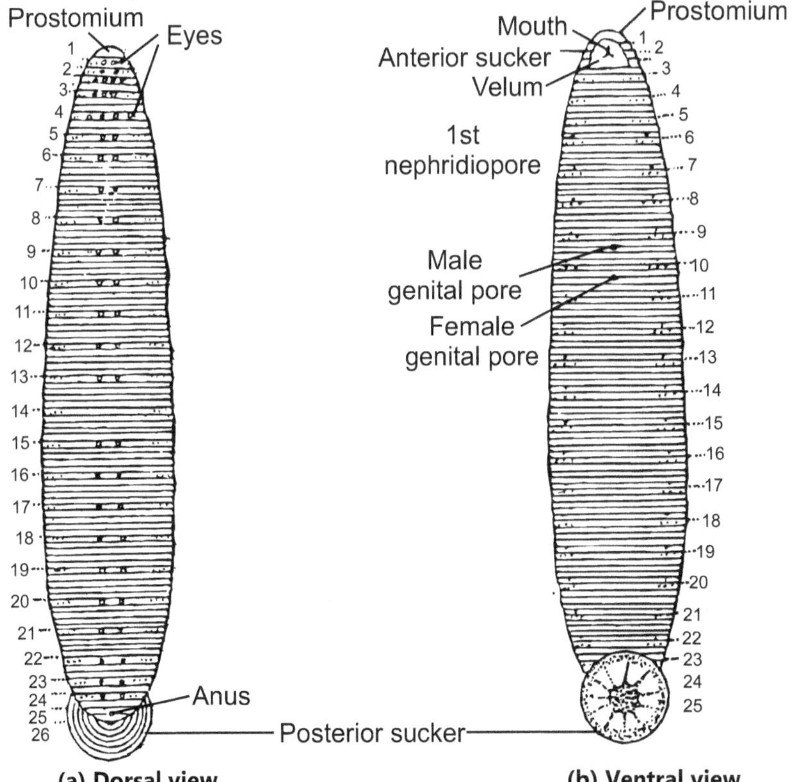

Fig. 2.5: *Hirudinaria.* **External features**

(5) Anterior and posterior ends of body are provided with suckers called cephalic and anal suckers respectively. They are useful for locomotion and attachment.

(6) Head is not distinct and bears only eyes.

(7) Coelomic (body) cavity is greatly obliterated and filled by botryoidal tissue.

(8) Body is very muscular and has got great power of contraction and extension.

(9) Sexes are united.

(10) Nephridia are excretory organs.

(11) Fertilization is internal and development is direct and it occurs in cocoon secreted by clitellum.

Examples: *Hirudinaria (leech), Acanthobdella, Placobdella, Hirudo.*

Sub-class-Brachiobdella:

(1) They are small (about 1 cm long) aquatic whitish animals.

(2) These animals are either commensals or parasites on crayfish.

(3) They are mostly found in the northern hemisphere.

(4) Different species attached to their hosts at different places on the body. For example, *Branchiobdella parasitica* attaches to the underside of the abdomen while *B. astaci* attaches to its hosts gills. Some forms are considered as parasitic feeding on the tissue of host but some forms are considered as commensals.

2.4 ECONOMIC IMPORTANCE OF ANNELIDA

Earthworms and leeches are the important members of the phylum Annelida. Both the animals are of economic importance to man. Directly or indirectly these animals are useful to man in the following ways.

1. As a bait and food: These animals are the major portion of the diet of birds, like robins, chickens, ducks, frogs, centipedes and small mammals. Earthworms are commonly used as bait for fishing all over the world. They are the best food of fish in aquaria. They are grown in the soil and used to feed aquarium fish and small laboratory animals. Earthworms are also used as a food in many parts of civilized world. Some leeches are used as bait for fish in place of earthworms. Like earthworms, leeches are said to be used as food by man.

2. In Agriculture: Earthworms are known as the friends of farmers because they plough the soil and make the soil fertile by adding castings in it. Although they may sometimes do damage to young and tender plants yet they are friends of gardners and farmers. The burrowing and swallowing earth below the surface increases fertility of soil. Due to burrows, air and moisture penetrates into porous soil and improve drainage and helps root growth. The castings of earthworms is a best quality manure for soil which increases the soil fertility. Earthworms can convert infertile land into fertile land. Considering the importance of earth worms now under biotechnology vermiculture has become popular and vermicompost manure production is helpful as biomanure for soil. This is the best manure as compared to chemical fertilizers.

3. Use in Medicine: In past earthworms were used as medicine. The medicines prepared from earthworms were used to cure stones in bladder, yellowness of jaundice, pyorrhaea, piles, gout, diarrhoea etc. Earthworms are also used in various fancy medicines in Japan, China and India.

4. Surgical Agents: In the past, Leeches were extensively used in surgery for blood letting under the mistaken belief that removal of bad blood may cure the disease. Leeches suck the blood from their prey in painless manner. Early in Greek medicines leeches were used for blood letting. Large number of leeches were used in the hospitals. The use of leeches in Ayurvedic medical practice in India is very ancient. Leeches were also used in the treatment of piles, tonsilitis and baldness. Hirudin extracted from leeches is used as haemolytic agent or anticoagulant in the modern hospitals, but now new and better anticoagulants have taken the place of hirudin.

5. Use in Laboratories: Earth worms and Leeches are of convenient size and used in dissection. Therefore, they are employed for class purposes in the zoological laboratories. They are also used in research purpose.

6. Harmful Worms: Earthworms sometimes become harmful. Exceptionally, the burrows made by earthworms may cause loss of water in irrigated lands. Their castings on slop land responsible for soil erosion. Some earthworms are an ectoparasites of frog and man. They burry in the carcasses of burried animals and bring disease causing pathogens to the surface where they infect other animals. The earthworms are said to serve as an intermediate hosts in the transmission of some parasites, such as tapeworm, gapeworm of chicken and the lung nematode of pigs.

Leeches are also known as transmitters of diseases because they act as vectors for pathogenic organisms. *Trypanosoma* of milk turtle is carried by vactor leech *Glossiphonia*. Padyfield Leech is carried of pathogens of rinder pest disease. Land leeches of Java are believed to transmit the flagellate causing gangrenous ulcers.

Leeches also act as pests because they are known for the blood sucking habits. Most of the leeches are parasitic, but others are scavengers. Few species of leeches feed on warm blooded vertebrates. They attack on human beings. They are found in large number in ponds and lakes causing nuisance as they are difficult to control. The land leeches are annoying pests. They are found commonly in Indian forests. These leeches attack horses, cattle and human being and cause serious loss of blood. The bites of leeches are followed by sores and ulcers which may lead permanent crippling. Leeches also attack on fishes in hatcheries and aquaria.

7. As predators: Leeches also act as predators because they destroy insect larvae, worms, molluscs and other leeches and invertebrates. They also attack on fish, frog, toads turtles, crocodiles, snakes, birds and mammals.

8. Vermiculture: Vermiculture essentially means the scientific culturing/rearing of earthworms. Vermiculture, Vermitechnology, Vermiculture bio-technology, Earthworm farming all deal with the culture and application of earthworms for various purposes viz. (1) As in the production of vermicompost (rich bio-fertilizer), (2) Earthworm protein as animal feed for livestock, (3) Some species for the treatment of industrial wastes. (4) Earthworms play a significant role in the biotic components of soil processes, which include turning and mixing of soils. The Green Revolution showed great promise, especially during the period between the 1960's and 1970's, during which food production increased substantially in India. This increase in food production, was due to a number of reasons i.e. more land was brought under agriculture, high external and chemical input based and ignores the fact that biological communities in the soil are essential for continuous soil fertility.

Among the 1800 known species of earthworm more than half belong to Megascolecides, is largest and widely distributed. It contains 30 Indian genera of which *Pheretima* is the largest.

Habit and Habitat: Earthworm is the best known, soil-inhabiting animal; occurs in diverse habitats and traced in forest, grassland, gardens orchards, plant-nursery and greenhouse. Organic material like manure, litter, composite, humus, effluents and kitchen drainage are highly attractive for some species, while others are hydrophilous and few live under snow and high mountains, hence identification of the various earthworm species on the basis of their micro-habitat is important to farmers, because such type of studies will provide guidelines for application of appropriate species which will suit to their agro-climatic condition and soil-texture. Ecological factors like pH, moisture content, salinity and temperature play an important role in the habitat of earthworm.

This soil aerator is an important bio-agent of the underground fauna, who is active day and night silently and secretely carrying on the tasking and laborious job of scavenger on the dead plant and animal remains.

Earthworm is a member of the soil biota (diverse assemblage of micro and macro organisms); and is master in mixing different soil components and production of surface and subsurface castings. Earthworm consumes the soil organic material and converts it into rich humus.

Useful Species of Earthworms:

All over the world, including the tropical and temperate regions many species of earthworms are tested for mass cultivation. These worms occur in diverse habitats i.e. organic materials like manure, compost, litter, humus, effluents and kitchen drainage etc. Earthworms are omnivorous but they mostly derive nutrition from dead organic matter, which generally does not occur abundantly in the soil. As a result, they are adapted to swallow large quantities of soil for extracting sufficient nourishment from it. Earthworms vary in size and shape. In India, some peregrine species like - *Microscolex phosphoreus* (Duges), *Dichogaster saliens* (Beddard) and *Bimastos parvus* (Eisen) are less than 20 mm long. While some endemic geophagous forms, such as *Drawida nilamburenis* (Bourne) and *Drawida grandis* (Bourne) may reach upto one meter in length. *Megascolides australis* (McCoy) from Australia is about 4 meters in length. The world's largest known worm *Microchaetus microchaetus* (Rapp) found in South Africa has a length of over 7 meters.

An African species, *Eudrilus eugeniae* and a European worm, *Eisenia fetida* are being cultured in several parts of the world. These species have also been transported to India and *Eudrilus eugeniae* is being cultured in South India for producing biofertilizer. The Indigenous compost worm species like *Perionyx excavatus, Perionyx sansibaricus* and *Dichogaster bolaui* could be used for vermicomposting. These species are surface dwellers and deeply pigmented and morphology of their alimentary canal (reduced or absence of typhlosole) indicated epigeic (surface dweller) way of their life. Certain litter dwelling species of *Perionyx, Amynthus* and

Megascolex, Lumbricus may also be used in the degradation of organic wastes.

In Maharashtra, species like *Pontoscolex, Corethrasus, Perionyx excavates, Perionyx sansibaricus, Pheretima elongate, Eudrigaster, Eudrilus eugeniae, Eugenia foetida, Pheretima posthuma* are widely being used. Since, these worms have do not only effect quick conversion, but also reproduce faster. The best choice for vermicomposting are two epigeic species - *Eudrilus eugeniae* and *Eisenia foetida.*

The following account describes some of the important species of earthworm and it is applicable more or less to other species also which are useful in vermiculture.

(a) ***Eudrilus eugeniae:*** It is one of the widely used exotic species in vermiculture. This cosmopolitan *Eudrilid* is originally from Africa and popularly called as the "African Night Crawler". This species is widely cultured and distributed by earthworm growers all over the world and has been found to be the best for vermicompositing and the production of protein meal. It has excellent growth and high conversion ratios. This species occur in orchard and natural farming with the temperature ranging from 19 - 22°C and soil depth 15 - 22 cm, with variable moisture content. This *Eudrilid* species have 8-setae per segment. Male pore is single on segment 17/18. They possess meganephridial nephridia.

In routine culture methods, optimum temperature for this species was found to be 20°C – 28°C. Hatching at 25°C, about 3 weeks cocoons were deposited. It can grow to a maximum weight of 2.5 g in 8-10 weeks.

(b) ***Eisenia foetida:*** This is a fast growing earthworm, with annual cocoon production almost 35 times as compared to that of *E. eugeniae.* The conversion ratios for this species are high and make it suitable for use in vermicomposting. The morphological features and other factors are more or less resemble with *E. eugeniae.*

(c) ***Perionyx sansibaricus*** **(Family-Megascolecidae):** It is one of the best known indigenous species for vermicomposting mostly occur in social forestry in the depth ranging from 3-8 cm, temperature 20-28°C but varied pH and at moisture content ranging from 20-40%.

These species have single pointed setae. Male pores are situated on 17th or 18th segment and female pores on 14th segment. The other species - *Perionyx excavatus* resembles with the *P. sansibaricils.* Both have been found to be good converters of organic waste and are better suited for the southern parts of the country where the summer temperature does not rise as high as in Central and North India. These species are purely used for the purpose of vermicomposting.

(d) ***Pontoscolex corethrurus:*** It is an endogeic (deep soil dweller) shallow burrowing worm, around 2 cm in length. This is a geo-phytophagous species and usually not used for vermicomposting.

(e) ***Eudichogaster species:*** It is one of Indian species of Michaelson. These found in monoculture vegetation and mixed farming at depth ranges from 10-30 cm and moisture content from 17-30%, temperature 19°C - 24°C.

It belongs to the family *Octochaetidae*. The pre-intestinal region is long with intestinal origin in or behind segment fifteen. Excretory system is mesonepheric.

(f) ***Polypheretima elongata*** **(Kingberg):** It is one of the most widely distributed species of the Megascolecid group. These species are found in monoculture farming at 20 cm depth, 25% moisture content and at the temperature 21°C.

It is characterised by the arrangement of numerous setae in a ring in each segment.

Of many species of earthworms, tested for mass cultivation all over the world, *Eisenia fetida, Eudrilus eugeniae and Perionyx excavatus* come in the order of preference for their ability to degrade the wastes. Their high biomass production may attain an increase of 40-90 times in a period of 3-6 months with adequate space and food.

Methods of Vermiculture :

Earthworms feed upon a variety of organic material and could be produced commerically for recycling biodegradable organic wastes, production of biofertilizers and animal protein for poultry and fish food. Vermiculture is feasible in suitable containers or specially designed boxes in small scale or shades in farm in large scale, since they are omnivorous, able to withstand environmental changes and resistant to many diseases. The technology is very simple and can easily be adapted in India, especially in rural areas. It is possible to culture worms both indoors and outdoors depending upon the local climatic conditions.

Some of the important guidelines for various culture methods are as follows:

(I) Culturing Techniques in Small Scale:

(1) The culture boxes or containers should be made up of light weight materials like plastic, wood, tin etc. (for indoor). Which could easily be carried from one place to another.

(2) The culture boxes or containers should be non-porous to minimise loss of moisture from culture medium.

(3) The size of the boxes/containers may vary according to the need. **Reynolds** (1977) used a specially designed wooden box which is more convenient and useful. It measures 50 cm in length, 35 cm

in width and 15-20 cm in depth. The bottom of the box is provided with a few holes of 50 mm in diameter.

(4) Plastic window screen is placed on the inside bottom with a burlop (or jute cloth) lining on top of the screened sides before the culture medium is added. This prevents the culture medium from sticking to the box and escape of worms through the holes but allows the excess of water to drain.

(5) Top of the box is covered with a burlop (or jute cloth) frame before inoculation of medium and earthworms.

The plastic tubs are more useful because these are more durable, lighter in weight and could easily be arranged one above the other in vertical rows on concrete shelves in limited space.

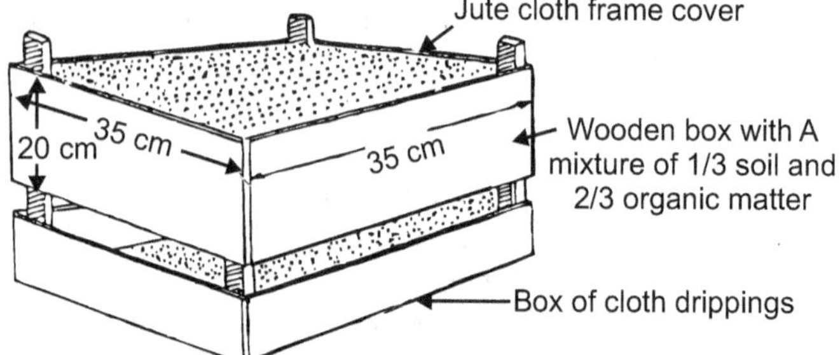

Fig. 2.5: A Design of Vermiculture Wooden Box

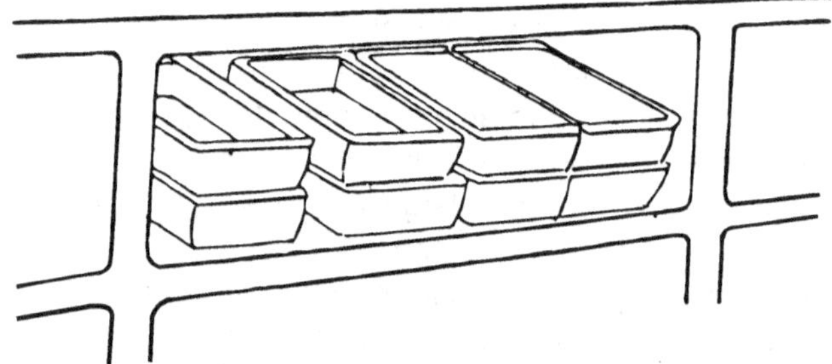

Fig. 2.6: A Diagrammatic Representation of Indoor Vermiculture in Vertical Rows of Plastic Tubs

(6) A mixture of 1/3 soil and 2/3 organic matter is to be more useful in culture containers. Beds in plastic or wooden boxes are prepared by spreading a sand layer of 2-4 cm in height over which another layer of equal thickness of soil is added. Organic matter is placed on one side of the container.

(7) Water is added to the culture medium so as to hold 25-30% of moisture.

Indoor cultures are preferably kept in a cool building at a temperature between 10°C and 15°C for lumbricids (e.g. *Eisenia fetida*) and about 20°C for tropical species (e.g. *Eudrilus eugeniae* and *Perionyx excavates*).

(8) Sources of common organic materials are : decayed leaves, hay, straw, rice or wheat bran, vegetable wastes, cow dung, poultry droppings, biogas sludge etc.

The following precautions should be taken during vermiculture:

(1) The culture medium must have sufficient organic material to avoid its formation into a soggy mass.

(2) Moisture of the medium should be maintained at required levels by sprinkling water regularly. Overwatering affects the culture adversely.

(3) Presence of a low watt light will prevent the worms from crawling out of　boxes.

(4) Outdoor cultures at places with low temperature in winters should be covered with suitable insulation materials like wheat straw, dry hay or weeds, manure, compost etc.

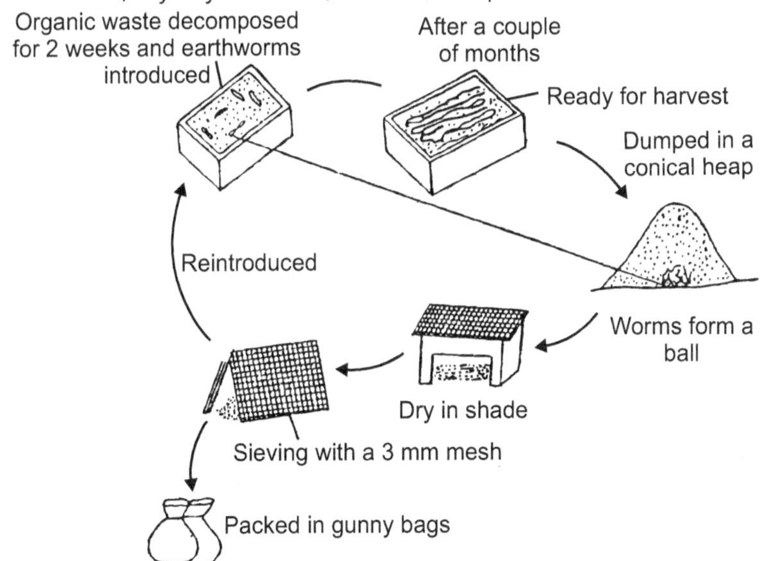

Fig. 2.7: One Vermicomposting Production Cycle

Large outdoor vermiculture beds may be established on waste lands. The culture bed is generally prepared with a bottom layer of 10 cm high gravel over which plastic window screen is placed with its edges raised upto 20 cm in height. A layer of 2-4 cm sand is laid over the window screen layer. A mixture of 1/3 soil and 2/3 organic matter is spread over the sand layer. The bed is slightly raised in the middle which allows drainage of excess of water on sides during the rains. The bottom layers of gravel and sand also help in maintaining the water content in the culture. The window screen prevents the escape of worms.

II. Culturing Technique at Farm and Production of Vermicompost:

Vermicomposting is the process of converting organic waste into vermicompost through the action of epigeic (surface dweller) earthworm species. Before the worms can ingest the waste, partial decomposition is essential. The mesophilic and thermophilic stages generate the heat in the compost pile; which destroy the earthworms. Once these stages are passed, the pile cools down and worms can be introduced.

Vermicompost is the cast or excreta of epigeic earthworm species which may be cultured on animal dung and other organic wastes.

Techniques:

(1) Containers: Any kind of container described earlier, will serve the purpose or container can be constructed in brick masonry and cement. Stone slabs can be used for the sides and bottom and all the joints are cemented. The constructed container should have a 1 m^3 volume. The dimensions are 1.6 m long, 1 m wide and 0.75 m high and contain around 6 to 7 thousand worms. The height of the waste pile should not be more than 0.6 m.

(2) Shelter: All the operations should be done under shade to prevent direct sunlight and rain.

(3) Feed mixture: Any organic waste like sugarcane thrash, coir pith, leaf litter, kitchen waste etc., can be mixed with dung in the ratio of around 1 : 8 and put in the container. If dung is not available, a little vermicompost from the previous harvest or even a little soil can be added to the waste. The waste must be allowed to decompose for 2 to 3 weeks. After the material cools, the worms can be introduced.

Production of Vermicompost: The worms feeding actively assimilate only 5 to 10% for their growth and the rest is excreted as loose granular rice shaped pellets. The castings are left on the surface.

Collection: The castings can be collected, once conversion is over, by observing pellets, the material can be dumped on the ground. Small conical heaps can be made and left for a few hours. The worms converge and form a ball at the base. The worms then taken out and introduced to the next feed lot. The casts are to be dried in the shade and then sieved through a 3 mm sieve to separate the cocoons and young ones which can be introduced to fresh culture beds.

To a common man earthworms are rather insignificant animals which generally come out on the soil surface during rains and serve as bait for angling but for farmers and agriculturist they are of atmost importance as they bring in the decomposition of organic matter in the soil.

Some of the most important factors in view of economic importance of vermiculture are as follows:

(1) Soil becomes loose and gets a greater capacity to retain air and moisture due to high densities of earthworms. Soils with earthworms drain water 4-10 times faster than soils without earthworms. Both these factors are important for growing crops.

(2) Burrowing habit of earthworms, leads tunneling in soil which aerate the soil and helps in increasing the air holding capacity of soil.

(3) Earthworms deposit their castings on the surface at night. In due course, they bring the sub-soil to the top and expose it to bacterial action. Bacteria helps in decomposition of cellulose.

(4) Earthworms also take the rich humus from soil surface to plant-roots and help in maintaining soil pH.

(5) Soil nitrogen is generally bound to organic complexes so it is not readily available to plants. This bound nitrogen is converted into available forms like ammonia, nitrates and nitrites as it passes through the digestive tract of earthworm thus earthworm castings are rich in nitrates, calcium, magnesium, phosphorus and potassium. This casting when added to the soil increases soil fertility and promptes plant growth thereby increasing crop production.

(6) Earthworm contains 50-70% of protein in dry weight basis and all essential amino acids. It is one of the important food materials for most of the animals.

(7) Earthworm tissues contain high amount of proteins which after proper processing could benefit as food to the live stock and aquaculture. Earthworms have been used as human food by Maoris in New Zealand and the natives of New Guineas as they are richer in arginine tryptophan and tyrosine.

(8) In Indian Unani system of Medicine, the external application of preparations made from the dried worms is used in treating wounds, piles, chronic boils, sorethroat, etc. and when taken internally for curing respiratory aliments, jaundice; rheumatic pains etc.

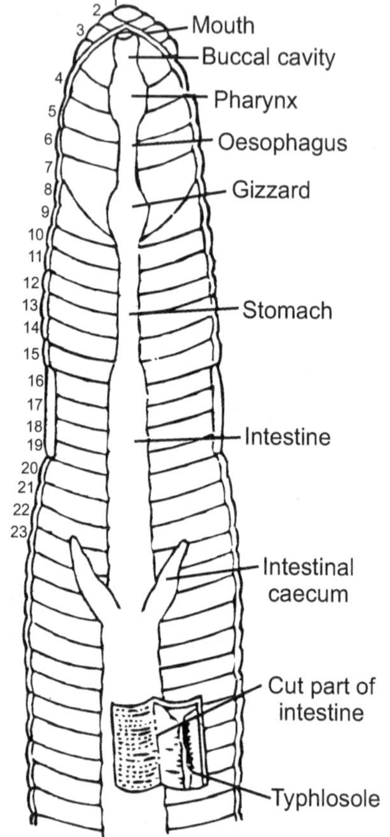

Fig. 2.8: Gut of the Earthworm

(9) Earthworms also play key role in environmental management. The sewage and city organic waste can be converted into vermicompost which acts as a good biofertilizer for plants.

(10) One of the significant advantages of vermicompost is the quick nutrient absorption by plants unlike other organic manures. This is because of the digestion of the organic matter by earthworms.

Vermicompost as Biofertilizer:

Vermiculture, the technology of producing rich biofertilizers and animal protein by using earthworms, has well established itself commercially in many developed countries. It has been estimated that one million worms can convert about 120 tonnes of organic wastes into biofertilizers in about one month's time. According to conservative estimates over 2,000 million tonnes of solid and liquid excreta of animals and human beings and another 200 million tonnes from crop straws are available as wastes annually in the country. Besides this there are vast quantities of domestic garbage and industrial wastes. Thus, production of biofertilizer through vermicomposting/ vermiculture has a bright future in India. Vermicomposting is the process of converting organic waste into vermicompost through the action of epigeic earthworm species. The earthworms consume the soil matter and convert it into humus within a short period of time and thereby increase the soil fertility within 24 hours. They can pass soil almost equivalent to their own weight through the 'alimentary canal excrete the soil as 'cast'. They have, therefore, rightly been called the nature's ploughman. Thus, the soil is being constantly and continuously turned over and over again by the worms i.e. organic matter is cycled and recycled, so that the humus formation is natural and a continuous process and enrichment of soil helps in agriculture and forestry. By planning and managing agricultural operations, supplemented with input from the earthworms, the farmers can get significant increase in crop yields. Countries like U.S.A., U.K. and Japan have realized this potential of earthworms and are therefore taking vermiculture quite seriously as an aid to farming.

As part of biotechnology, vermiculture has attracted attention, since it is an entirely natural process which maintains the environmental balance and leaves no adverse effects.

According to **Bhole**, the soil with worm casts, in comparison to soil without the worms, has 5 times more nitrogen, 7 times more phosphorus, 11 times more potassium, 2 times more magnesium and calcium each; along with other trace elements and soil nutrients are soluble in water and readily available to the root systems.

Increase in the organic wastes mainly due to growth of human population, agricultural and industry is a global problem and a serious constrain in the maintenance of a clean and healthy environment; but due to benefit of vermiculture, the earthworms should be considered nature's most useful converters of these waste products. Experiments with worms have shown that the recycling the utilisable organic wastes arising out of household garbage, city refuge, sweage sludge and paper, food and wood industries.

Vermitechnology has the following Applications:

(1) Use of epigeic earthworms in the production of vermicompost. This technology can be used for urban and rural waste recycling for conversion of organic wastes to manure.

(2) Vermicompost as a manure in agriculture or as inoculum for improving and maintaining soil fertility.

(3) Earthworm protein as animal feed for livestock. (Protein meal can be prepared from many earthworm species and used as livestock. This meal content 60% protein, 7-10% fat, 8 to 10% ash and 0.55% calcium, 1% phosphorus).

(4) Earthworm species for the treatment of industrial wastes.

(5) Earthworms play a significant role in the biotic components of soil processes, which include turning and mixing of soils.

(6) Earthworm (*Eugenia foetida, Eudrilus eugeniae*) species are used for treatment of different types of industrial effluents.

POINTS TO REMEMBER

- Annelids are aquatic and terrestrial animals.
- Body is metamerically segmented.
- Setae or chaetae are the locomotary organs.
- They are triploblastic and true coelomate animals.
- Closed type of circulatory system.
- Nephridia are the excretory organs.
- Nervous system consists of brain, nerve cord and segmential ganglia.
- Sexes may be separate or united.
- The Phylum Annelida has three classes, such as Polychaeta, Oligochaeta and Hirudina.
- Earthworm and Leeches are of economic importance to man.
- Earthworms are useful in agriculture, as a food, medicine and in laboratories.
- Leeches were used as a surgical agents and in Ayurveda.
- Many important species of earthworms are useful in vermicomposting.
- Vermicompost is used as a biofertilizer.
- Vermitechnology has several applications.

EXERCISE

1. Give the distinctive characters of Polychaeta.
2. Give the distinctive characters of Oligochaeta.
3. Give the distinctive characters of Hirudinea.
4. Give an account of general characters and outline of classification of Phylum Annelida with suitable examples.
5. Give an account of economic importance of Annelida.
6. Describe various methods of vermiculture.
7. Write short notes on :
    (a) Economic importance of Annelida.
    (b) Economic importance of Vermiculture.
    (c) Vermicomposting.
    (d) Vermiculture.
    (e) Vermicompost as a biofertilizer.

3
CHAPTER

PHYLUM - ARTHROPODA

CONTENTS

3.1 INTRODUCTION TO PHYLUM ARTHROPODA

Arthropoda means 'jointed legs' (Greek; Arthros = jointed; Podos = foot.) Therefore, jointed legs is the most important characteristic structure of this phylum.

3.2 SALIENT FEATURES OF PHYLUM ARTHROPODA

1. The animals are bilaterally symmetrical and metamerically segmented.

2. The body is covered externally by thick, tough, non-living, organic chitinous exoskeleton.

3. Exoskeleton is cast off periodically and this process is called ecdysis or moulting. It is essential for the growth of the animal.

4. The appendages are jointed and they are modified as gills, jaws, legs etc. for various functions.

5. They are triploblastic animals but true coelom is reduced in adult and represented by excretory and reproductive organs.

6. Respiration may occur throughout the body surface but generally takes place by special structures like gills, tracheae and book lungs.

(3.1)

7.   Circulatory system is of open type i.e. the blood flows in spaces or sinuses comprising haemocoel instead of blood vessels.
8.   True nephridia are absent. Excretion occurs by green glands or by malpighian tubules.
9.   Nervous system is of annelidan type.
10. Compound eyes are present.
11. Sexes are usually separate and often there is distinct sexual dimorphism.
12. Fertilization is generally internal. Development includes larval forms which undergo varying degree of metamorphosis to become adults. Parental care is often well marked.

3.3 CLASSIFICATION OF PHYLUM ARTHROPODA

Class-Arachnida:
(1)   Mostly terrestrial, solitary, air-breathing arthropods.
(2)   Free living, predaceous or parasitic.
(3)   No head, body divisible into prosoma (cephalothorax) and opisthosoma (abdomen).
(4)   The cephalothorax bears simple and sessile eyes and six pairs of joined appendages. Of these one pair is of chelicerae, one pair pedipalpi and four pairs of walking legs.
(5)   Antennae and mandibles are absent.
(6)   Abdomen is without appendages.
(7)   Cuticle is with sensory hairs or scales.
(8)   Mouth parts and digestive tract are adapted for sucking.
(9)   Respiration by tracheae or book-lungs or both. In aquatic forms, respiration is by cutaneous or by book gills.
(10) Excretion takes place by Malpighian tubules or coxal glands or both.
(11) Sexes are separate and sexual dimorphism is inconspicuous. Males are smaller than females.
(12) Fertilization is internal. Oviparous and few viviparous. Development is direct.

Examples: *Palamnaeus, Buthus* (Scorpions), Spiders, Ticks and Mites.

Class-Crustacea:
(1)   They are aquatic marine and freshwater.
(2)   Some are parasitic or sedentary.
(3)   They have variously shaped bodies with three divisions; head, thorax and abdomen.

(4)   Exoskeleton is made up of thick protective, chitinous cuticle strengthened by impregnation with calcium salts, hence crustacea.

(5)   The thorax may fuse partly or wholly with head to form cephalothorax, which is partly or entirely covered by exoskeletal shield called carapace from dorsal side.

(6)   Head bears pair of stalked compound eyes.

(7)   Jointed appendages vary in number but there are two pairs of antennae, one pair of mandibles and two pairs of maxillae on head and one pair of appendages on each segment of thorax and abdomen.

(8)   First pair of antennae is uniramous but remaining all appendages are biramous. The appendages are variously modified as jaws, legs, fins, gills or accessory reproductive organs.

(9)   Respiration by gills or pseudotracheae or by body surface.

(10) Excretion by green glands. Malphigian tubules are absent.

(11) Mostly unisexual except cirripedia. Sexual dimorphism is common. Generally, females carry the eggs.

(12) Some groups like **Brachiopoda** and **Ostracoda** show parthenogenesis.

(13) Development includes more or less of metamorphosis and is accompanied by free living larva nauplius.

Examples: *Lepas, Daphnia, Cypris, Cyclops, Palaemon,* **Lobster, Crab.**

Class-Myriapoda:

(1)   They are terrestrial, air breathing.

(2)   Body elongated vermiform with distinct head and trunk has number of segments.

(3)   Many jointed pair of antennae are present, pair of mandibles and maxillae are present.

(4)   The body segment bears one or two pairs of jointed legs.

(5)   Respiration by spiracles and tracheae.

(6)   Malpighian tubules are excretory organs.

(7)   Sexes are separate, single gonad with paired gonoducts.

Examples: *Spirobolus* **(Millipede),** *Scolopendra* **(Centipede).**

Class-Insecta:

(1)   Mostly terrestrial, aerial, air breathing animal. Rarely aquatic.

(2)　Body is divided into three distinct regions - head, thorax and abdomen.

(3)　The head is formed by fusion of six segments and bears a pair of antennae, a pair of compound eyes, few simple eyes or ocelli, a pair of mandibles, two pairs of maxillae.

(4)　The mouth parts are variously adapted for chewing, biting, piercing, sucking, lapping or siphoning etc.

(5)　The thorax consists of three segments and bears three pairs of jointed legs (hence hexapoda), two or one pair of wings present dorsolaterally. Some insects are wingless.

(6)　The abdomen consists of 7 to 11 segments and lacks appendages but genitalia are present at the end.

(7)　Salivary glands are present but liver is absent.

(8)　Respiration by tracheae opening or by paired segmental spiracles.

(9)　Heart is tubular muscular and 8 chambered situated in the abdomen.

(10)　Excretion occurs by two to many Malpighian tubules attached to the anterior end of the hind gut.

(11)　Nervous system is of the annelidan type.

(12)　Sense organs are well developed.

(13)　Sexes are separate, sexual dimorphism is distinct.

(14)　Gonads are composed of numerous tubules or follicles.

(15)　Fertilization is internal. Development usually occurs with some sort of metamorphosis. Parthenogenesis takes place in some forms.

Examples: ***Mantis, Gryllotalpa*, Dragonfly, Grass hopper, *Pediculus* (Louse), Butterfly, Moth, Honey bee, Beetle.**

Habits and Habitat and Distribution of Arthropoda:

Arthoropods are found all over the globe in diverse habitats, wherever life is possible on the earth. They are seen over high mountain ranges 20,000 feet above the sea level. They are also found at the deep sea bottoms. Arthropods occur in freshwater lakes, streams, ponds, sulphur springs, hotsprings and even pools of petroleum. They are also inhabitant of dry deserts and undergrounds. They live as ecto or endoparasites in other animals and plants including man. They can fly in the air, swim, hop, crawl. Many species of the arthropods are gregarious, while some are social insects which show well marked division of labour among the members of different castes.

Most of the aquatic forms are distributed in the lakes, ponds, stream and rivers of the world. Marine forms are also seen in all the oceans. The locust insects occur all over the world. Desert locust found in North Africa to North India. Bombay locust is confined to India only. American species are found in America. Many species of Arthropoda are abundant in tropical region. Cockroaches flourish chiefly in tropical damp forests, but they are throughout the world except for the polar regions. Some species (*Blatta*) native of tropical Asia. Some are native of central Europe. *Periplaneta americana* is originally from tropical America (*Mexico*) has travelled with man to all parts of world.

3.4 MOUTH PARTS IN INSECTS

The mandibles and maxillae, along with the labrum and hypopharynx form the mouth parts useful for feeding insects show diverse modes of feeding and hence mouth parts are variously modified in different insect groups to suit their modes of feeding. The study of mouth parts of insects is important to control the insect pests.

Following are the main modifications occur in different insects.

1.	Biting and chewing mouth parts.

2.	Chewing and lapping mouth parts.

3.	Piercing and sucking mouth parts.

4.	Sponging and lapping mouth parts

5.	Siphoning mouth parts.

**1.	Biting and Chewing Mouth Parts:**

These are the primitive mouth parts and from them all other types of mouth parts have evolved. They are also called as mandibulate type of mouth parts and are meant for pinching off, chewing up and swallowing the pieces of plant and animal tissues, chewing type of mouth parts occur in Cockroaches, Grasshoppers, Crickets, Termites, Beetles, Dragon flies and Silver fish. These mouth parts consist of following parts:

**(1)	Labrum:** It is broad flap having variable shape and size hanging from the clypeus in front of other mouth parts. It forms the roof of the mouth cavity and show slight upward and downward movement. It has on its posterior (ventral) side a membranous swollen area, the *epipharynx* which bears taste buds.

(2) Mandibles: Their are pair of mandibles which are heavily sclerotized, triangular structures lying just behind the labrum. Each mandible bears teeth like denticles, which interlock while capturing the prey. The mandibles are moved by a pair of abductor and adductor muscles. The mandibles are useful for biting and chewing the food.

(3) Maxillae: The maxillae are paired situated on the sides of the mouth behind the mandible. They are regarded as the appendages of 5th head segment. Each maxilla consists of a basal two pieces, a proximal *cardo* and *distal stipes*. This portion is called *protopodite*. The inner *endopodite* has also two segments, medial jaw like lacinia and lateral blunt *galea*. The outer *exopodite* or *palp* of usually 5 segmented called maxillary palp functioning as a tactile organ. The lacinia and galea which bear taste buds so they taste the quality of food. The maxillae are useful for holding the food material.

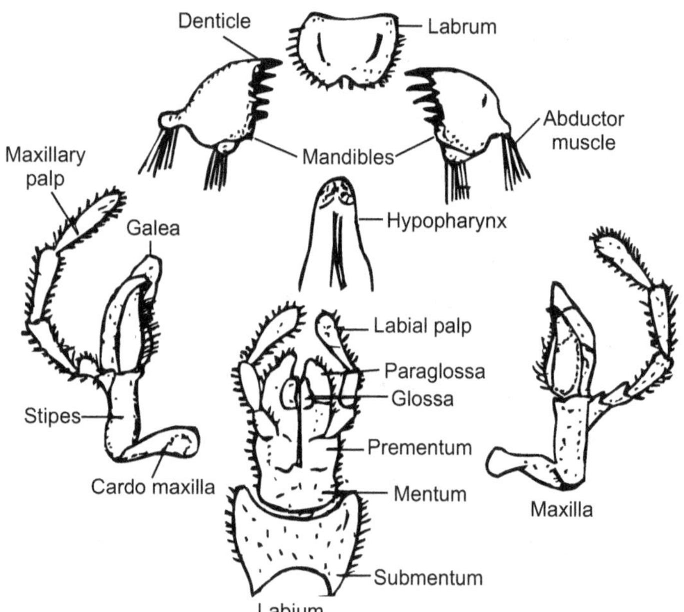

Fig. 3.1: Biting and chewing (Mandibulate) type of mouth parts

(4) Labium of lower lip: This is also called 2nd pair of maxillae, basically fused to form a broad plate divided by a transverse suture into two parts, the basal *postmentum* and the distal *prementum*. The postmentum may be divided further into a basal *submentum* and distal *mentum*. The prementum bears a pair of usually three jointed palps called labial palps laterally. In between the palp lie two *glossae* and *paraglossae*.

They are commonly called lingula. The labial palps are sensory and help in testing the quality of food as they contain chemoreceptors. The glossae and paraglossae together prevent the loss of food particles while feeding. They also aid in pushing the masticated food into the preoral cavity. These organs are formed from the 6th head segment.

(5) Hypopharynx: It is a short, median, tongue like appendage in the mouth cavity between the labium and mandibles. It is not muscular nor glandular but supported by few narrow sclerites. The common salivary duct opens into the base of the hypopharynx.

2. Chewing and Lapping Mouthparts:

This type of mouth parts found in some Hymenoptera (bees and wasps), particularly in worker bees. They are used for collecting nectar and pollen of flower and moulding wax. These mouth parts serve for both biting and licking. The labrum lies beneath the clypeus and fleshy epipharynx projects below the labrum. The mandibles are smooth and spatulate. The worker bees use them in building the cells of the hive. The maxillae lack laciniae have vestigial palps and bear long blade like galeae. The labium has elongated palps, reduced paraglossae and elongated curved glossae united to form a retractile tongue bearing a *honey-spoon* or *labellum* at the distal free end.

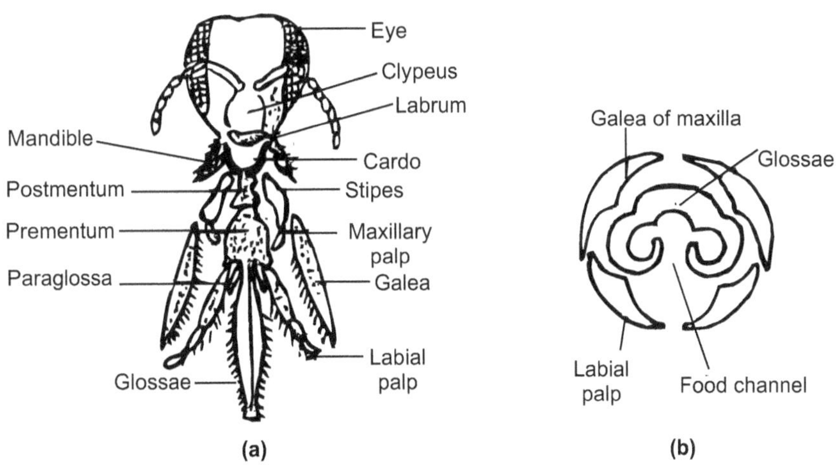

Fig. 3.2: Labellum (Honey spoon)

A – Mouth parts of worker Honey-bee; B – T. S. of mouth parts

While feeding, galeae and labial palps are brought close together forming a long hollow proboscis which can be inserted deeply into the corolla of a flower. Nectar is drawn up the proboscis by back and forth movement of the tongue in it.

3. Piercing and Sucking Mouth parts:

These mouth parts occur in bugs, aphids, scale insects, mosquitoes, fleas and some flies. They are used for sucking the sap or blood by piercing into plant or animal tissue.

(1) Mouth parts of Mosquito: These mouth parts are used for piercing and sucking the blood from host. The labrum, mandibles, maxillae and hypopharynx form long, pointed *stylets* for piercing the host skin. The labrum grooved ventrally. The hypopharynx encloses the salivary duct. The labrum, hypopharynx by getting closely applied together enclose the food channel for drawing blood. The labium forms a long hairy, dorsally

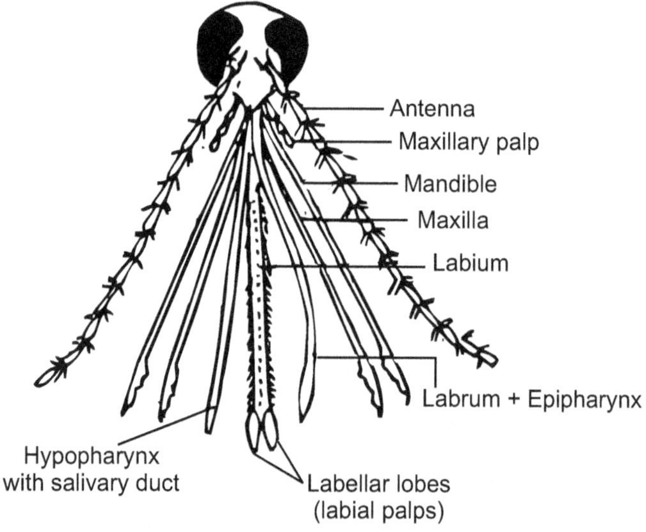

(A) Mouth parts of female *Culex*

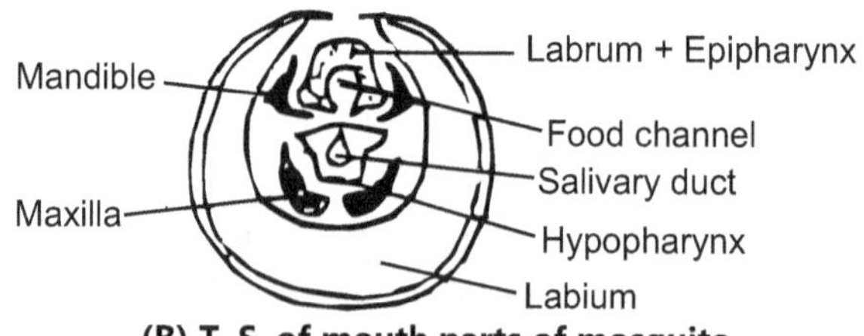

(B) T. S. of mouth parts of mosquito

Fig. 3.3: Piercing and sucking mouth parts

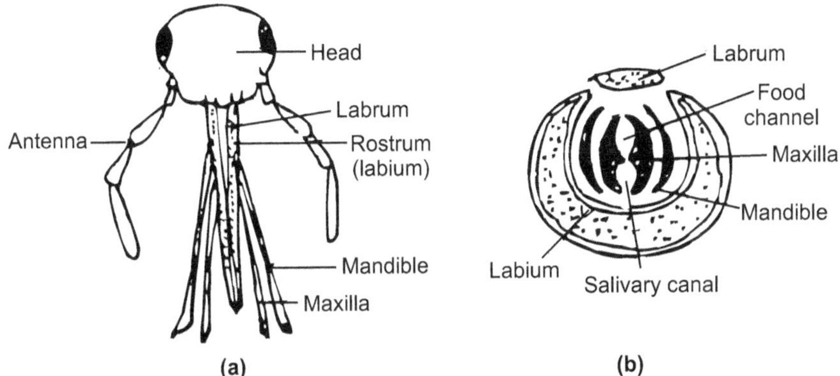

**Fig. 3.4: (a) Piercing and sucking mouth parts of bug,
(b) T. S. of mouth parts**

grooved *proboscis* tipped with a pair of sensory *labellar lobes* representing the labial palps. It serves as a sheath for the stylets at rest but flods up as the stylets pierce the host tissue at the time of feeding. The mandibles are absent in the male. Maxillary palps are short and 3 jointed.

(2) **Mouth Parts of Bug:** In the bugs, the labium forms usually a 3 jointed rostrum or proboscis. The mandibles and maxillae form four long, needle-like piercing stylets. The proboscis encloses the stylets. The hypopharynx is short lobe within the base of the proboscis. The inner stylets in the proboscis i.e. the maxillae fit together in such a way as to form the food and salivary channels. Maxillary and labial palps are absent.

4. **Sponging or Lapping Mouth parts:**

These mouth parts are common in houseflies, blow flies, fruitflies. They are modified for sucking up the liquid food by salivary secretion. The mandibles are absent, while the maxillae are represented only by two maxillary palps each made up of a single piece. The labium is greatly modified to form so-called *proboscis* which consists of three parts.

(i)   The proximal *rostrum* bearing maxillary palps.

(ii)   The middle haustellum with a mid-dorsal groove serving as the food passage and ventral plate like structure called theca and

(iii)   The distal labellum (oral disc) or sucker. The labellum has two lobes or *labellae* on which innersides contain many cylindrical tubes called *pseudo-tracheae*. They converge into the mouth that leads into the food channel. The pseudo-tracheae soak the fluid food, which is then passed on to the food channel from where it is sucked up by muscular pharynx. The food channel is formed by labrum, epipharynx and hypopharynx.

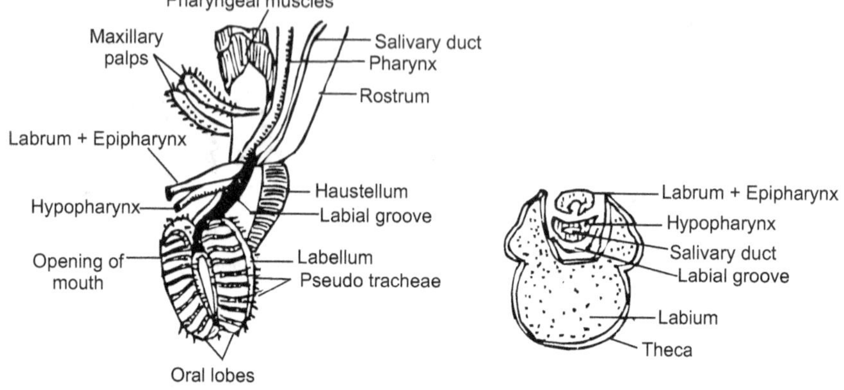

(A) – Proboscis　　　B – T.S. of Haustellum

Fig. 3.5: Mouth parts of Housefly (Siphoning type)

5. Siphoning Mouth parts :

These mouth parts occur in butterflies and moths and they are modified for sucking the fluid nectar from the flowers and fruits. The mandibles and, the labium are very much reduced. The labrum is merely a narrow transverse band. The maxillary palps are vestigial, but the galeae are very long and deeply grooved medially. When applied together the two galeae enclose a food channel and form as prominent proboscis. Which is the main siphoning tube. When it is not in use, is coiled like a watch spring under the head. At the time of feeding the proboscis remains uncoiled, inserted into flowers, fluid nectar is sucked by muscular pharynx which is present at the base of proboscis.

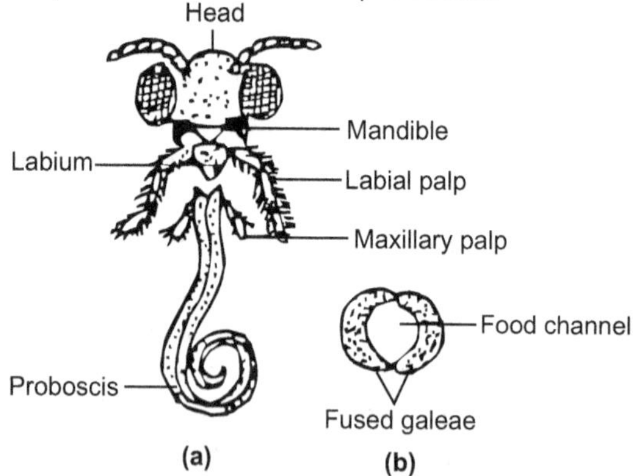

(a)　　　　(b)

Fig. 3.6: Mouth parts of Butterfly. (a) Front view; (b) T. S. of Proboscis

The labium is reduced to a triangular plate bearing well-developed, 3-jointed labial palps. There are no hypopharynx and salivary channel are lacking.

3.5 ECONOMIC IMPORTANCE OF ARTHROPODA

3.5.1 Useful Insects : Honey Bee, Lac Insect, Silkworm

Honey Bee:

Honey bee (*Apis indica*) is very useful insect for man because it gives honey, wax and it helps in pollination. Honey is social insect living in colonies of 50,000 or more individuals. They feed on pollens and nectar of the flower. They live in highly organised colony in which different castes are present like queen, drones (males) and worker bees or sterile females. There is good co-ordination among the members of colony and labour of division. There are different species of honey bees. There are 90 percent worker bees in the colony.

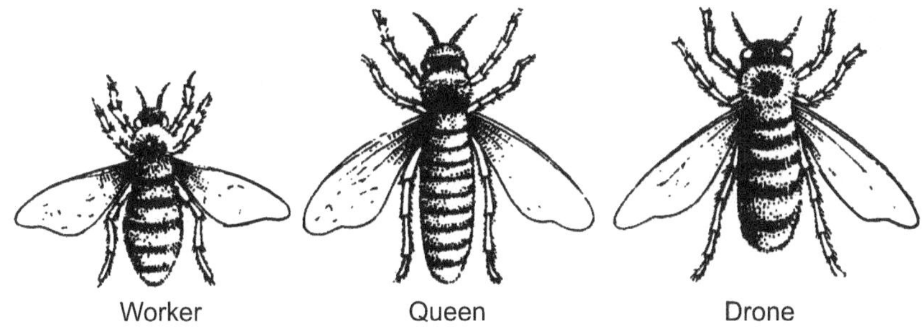

Fig. 3.7: Castes of Honey Bee

Workers attend all duties of food collection, bringing nectar, secreting wax, tending the young, building and cleaning the comb. Their mouth parts are modified for collecting nectar and moulding wax, the epidermis of abdomen for secreting wax and their legs for collecting pollen. In queens and drones, the mouth parts are shorter because they do not collect nectar, their epidermis do not contain wax glands and modifications of metathoracic legs are absent. Queen is large with longer abdomen. It does not perform function of nest making and pollen gathering. It has no wax glands and modified legs for collection of pollens. It has sting to combat a rival queen. The role of queen is to produce large number of eggs. Drone or male is larger and stouter than worker. It has no sting and they are formed from unfertilized eggs. In the hive bees live. Comb is made by workers with the help of wax secreted by them. Birds, moths and wasps are the enemies of bees. Man has domesticated honey bees to get honey and wax. Honey bees bees also play very important role in pollination. Honey collect the nectar and bring to their hive and regurgitate in the hive. Chemical changes occur due to mixing with saliva. Sucrose is hydrolysed to glucose, levulose and fructose which are more readily assimilable by man. Honey is stored in the cells of hive. The cell is capped with wax plug. It is the food of adults and larval bees. Honey is used as the chief source of natural sweet in preparing candies, cakes and bread etc. It forms a very important food for patients of diabetes. It is powerful tonic as it contains Vitamin B, Riboflavin, Vitamin C. Bee wax is used in cosmetics and toilet goods. The wax is also used in preparing candles, shaving creams, cold creams, cosmetics, polishes, casting of models, carbon paper, crayons, electrical and other products. In agriculture they play very important role of pollinators and hence there is good yield of fruits and grains. They have great value in agriculture.

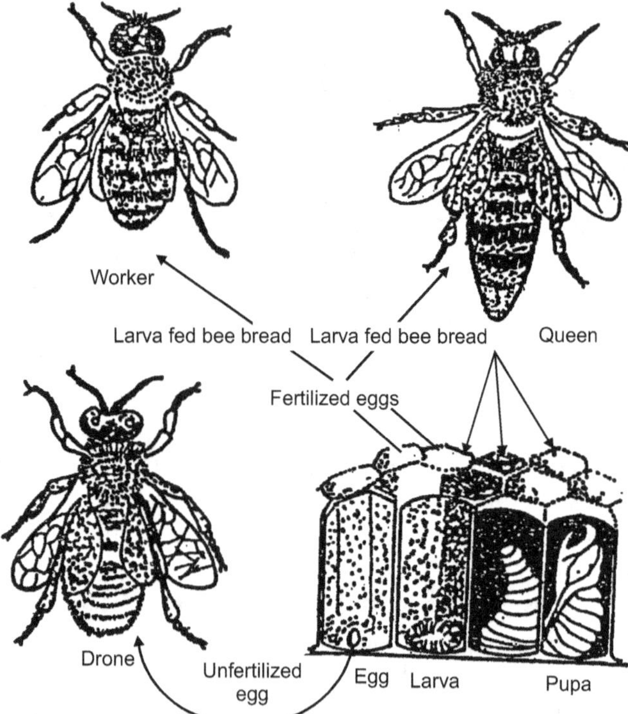

Fig. 3.8: Life Cycle of Honey Bee

Silkworm: The silkworm or silkmoth is an insect which belongs to the order Lepidoptera of the class insecta and phylum Arthropoda. The true silk is obtained from the cocoons of insects popularly known as silkworm. The rearing of silkworms for the production of raw silk is known as *sericulture*.

Silk and silk garments have always been the first choice of the members of the Royal families and other associated with them in India. It is one amongst the oldest business being conducted in our country and a large number of families are associated with the production of silk in India. It is a fact that it has never been a native of India but today the Indian silk is on the top both in quality and quantity.

In India first of all Lefroy (1904-05) started investigations on silk worm and sericulture at Pusa Institute, New Delhi. Today it is well known cottage industry achieving continuous progress through improved Technology. India has silk producing centres in Assam, Bengal, Chennai, Mysore, Karnataka and Jammu and Kashmir. India is second in the world production of Tassar silk.

The modern silk industry in India has grown to meet the domestic rather than export requirements. Further, this industry provides employment opportunities to about 60 lac people in India. It plays key role in the uplift of rural economy besides earning considerable foreign exchange.

Species of silkworms: Silk is prepared by the species belonging to family *Bombycidae* and *Saturniidae*. *Bombyx mori* is well known silkworm belongs to *Bombycidae* while *Eri, Tasar* and *Muga* moths belong to the family *saturniidae*.

1. **Mulberry silkworm (*Bombyx mori*):** It is the native of China and now this silkworm has been introduced in all the sillk producing countries like Japan, India, Korea, Italy, France. Its natural food is mulberry leaf so it is called mulberry silk. The colour of silk produced by this silkworm is white or yellow.

2. **Tassar silkworm (*Antheraea paphia*):** It is commonly known as Japanese oak silkworm and also reared on large scale in Japan. It produces coarse silk. These species feeding on Sal, Asan, Arjun and Ber leaves. These can not be domesticated and are collected from the forest and wild shrubs and trees.

3. **Muga silkworm (*A. assamensis*):** It is semidomesticated silkworm. The native place of this species is Assam, where now it has become good source of cottage industry. The silk is produced by this species is called muga silk which is of not good quality.

4. **Eri silkworm (*Attacus ricini*):** Feeding upon the leaves of castor is completely domesticated and produce deep brown coloured silk. It is dull in colour but stronger than cotton threads. The rate of silk production is much slow and quality is of lower grade.

5. **Oak silkworm (*A. pernyi*):** It is known as chinese oak silk worm which yields silk is large quantity. The colour of silk is pale buff.

6. **Glant silkworm (*Attacus altas*):** It is found in India and Malaysia, which is the largest of all the living insects.

Mulberry Silkworms:

Bombyx mori or the Mulberry silkworm is completely domesticated insect and is never found wild, which has been exploited for over 4000 years. All the strains reared at present belong to the species *B. mori* which is originated from the original Mandarina silkworms, known as *Bombyx mandarina* Moore.

The adult moths of *B. mori* seldom eat and are primarily concerned with reproduction. Their larvae are voracious eaters. They feed on the leaves of mulberry trees. Some moths are single brooded or *univoltine* and others are many brooded or *multivoltine*. Owing to domestication, a large number of strains have evolved out, which produce cocoons of various shapes, sizes, weights and colours ranging from white to yellow. Only one generation is produced in one year by worms in Europe and other countries where the length of winters far exceeds the duration of summers. Some strains pass through two to seven broods and are cultivated in warm climates. In South India particularly Mysore, Coimbatore and Salem, a strain which produces several generations, extensively utilized to produce silk.

Bombyx mori produces cocoons with continuous silk filament and therefore can be industrially reeled to produce raw silk. The mulberry silkworm may be further classified and identified as of Japanese, Chinese, European or Indian origin based on geographical distribution or as univoltine, bivoltine and multivoltine depending upon the number of generations produced in a year under natural conditions.

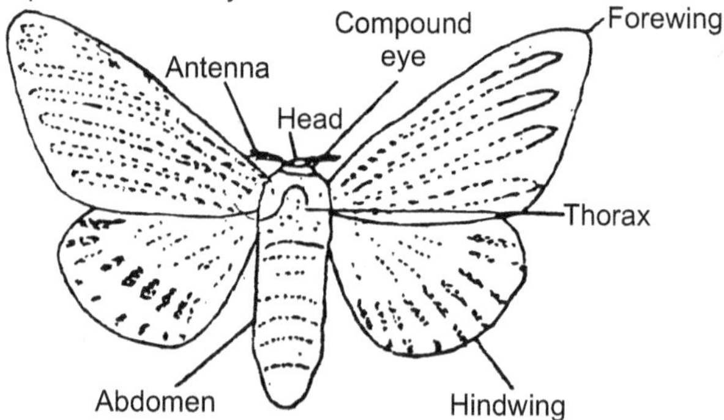

Fig. 3.9: Moth (*B. mori*)

Eri Silkworms:

Eri silkworms *Samia cynthia* ricini (Hutt) belongs to family saturniidae. The local name eri derives from its primary food plant 'era' (castors) so commonly known as castor silkworm is a domesticated and reared on castor oil plant leaves so as to produce a white or brick-red silk popularly known as *eri silk*. The filament of the cocoons spun by these worms is neither continuous nor uniform in thickness, the cocoons cannot be properly reeled and therefore, the moths are allowed to emerge and the pierced cocoons are used for spinning purposes to produce the eri silk yarn.

The distrubution of eri silkworm in north eastern India is mostly confined to the Brahmaputra valley and the surrounding areas extending to the foot hills of Meghalaya, Mizoram, Nagaland, Manipur and Arunachal Pradesh upto about 3000 feet altitude. Manipur are predominantly eri growing areas of this region. Eri silkworm is polyphagous in nature. Its food plants are abundantly found in natural forests in the plains and hilly areas of north eastern India. However, castor (*Ricinus communist*: Linn) is the major food plant; while Kesseru (*Heteropanax fragrans* Seem), Tapioca (*Manihot stilissima* Pohl) and barkesseru (*Ailanthus excelsa*) are the secondary food plants of eri silkworm.

Both male and female have brown, black and green coloured wings with white crescent markings and woolywhite abdomen. The male is smaller than female bearing bushy antennae and narrower abdomen.

This species is domesticated and multivoltine in nature having six generations in a year. *S. ricini* exhibits several larval strains viz: plain, spotted, semizebra and zebra based on larval markings and white, blue and green based on larval body colour. These strains cross breed naturally.

Eri-culture is a traditional practice among the Indo-Mongolid and Tibet Burman sub-tribes of this region. This region accounts for more than 90% of the the total eri-silk production. From time immemorial, The eri-culture remained apart of tribals way of life. Hence, the eri-fabric is commonly known as 'poor man's wool' or 'poor man's silk'.

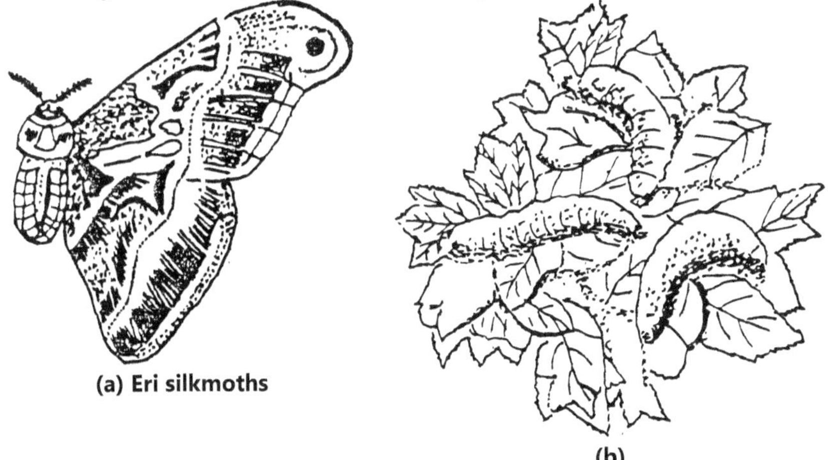

(a) Eri silkmoths

(b)

Fig. 3.10: Eri silkmoth and Silkworm larvae of *Attacus ricini*

Tasar Silkworms:

The tasar silkworms belong to the genus *Antheraea* and they are all wild silkworms. There are many varieties such as the Chinese tasar silkworm - *A. pernyi* G. The Indian tasar silkworm - *A. mylitta.* The Japanese tasar silkworm - *A. yamamai* Guerin.

Tasar silk occupies the third position; next to mulberry and eri silk. China is the biggest tasar silk producer of the world; followed by India. In the North Korean provinces bordering China, a small quantity of tasar silk is also produced in Japan.

Traditional tasar silk of India of is the one produced by the tropical tasar silkworm, *Antheraea mylitta* its distribution extends along the tropical forest belt of India starting from West Bengal in the east, extending upto Karnataka in the South-East through Bihar, Orissa, Uttar Pradesh, Madhya Pradesh, Andhra Pradesh and Maharashtra. Extending along the Himalayas right from Arunachal Pradesh in the east Jammu and Kashmir in the west using the temperature breeds of Tasar silkworm like *A. pernyi.*

Tasar silkworm is polyphagous, feeding on about two dozen host plants belonging from different families, genera and species but important are:

(a) *Terminalia* (Vernacular asan, ain)

(b) *Terminalia arjuna* (V; sal) (V. Arjun)

(c) *Shorea robusta* (V; sal)

(d) *Lagerstroemia sp.* (V, Sidha)

(e) *Ziziphus jujuba* (V. bar)

Unlike other silkworms, tasar silk worms being entirely wild, have to be fully reared on host plants. Tasar moths are fairly large insects, females being larger and yellowish brown in colour while males are smaller and brick red in colour but both having eye spots on their wings. Antennae of male are busy (branched) and narrowed abdomen.

The worms are either uni or bivoltine and their cocoons like the mulberry silkworm cocoons can be reeled into raw silk.

Fig. 3.11: Tasar silk moth Antherea paphia (after Lefroy)

Muga Silkworms:

The muga silkworms (*Antheraea assama*) also belong to the same genus as tasar worms but produce an unusual lustrous golden-yellow silk thread which is very attractive and strong.

The primary food plants of *A. assama* are Som (*Muchilus bombycina*) and Soalu (*Litsaea polyantha*) leaves which are Assammese names.

Muga silkworms are wild in nature, *muga* moths, (muga is an Assamese word meaning brown or amber) are distributed from western Himalayas to Nagaland, cachar districts of Assam of south Tripura. But the sericulture practice is confined to the Brahmaputra valley of Assam and Foothills of East Garo hills of Meghalaya. Ideal temperature for muga silkworm growth is 24-30°C and humidity 75-85%.

The wings and body of the male moth are copper brown to dark brown while female is yellowish to brown in colour. Both pairs of wings bearing eye spots. Males are smaller in size, slender abdomen, with bushy antennae. Assam and Meghalaya account for more than 90% of the production of muga silk. This silk is a traditional costume during the marriage ceremonies and festive occasions; the ladies garments 'Mekhala and Chadhar' made of muga silk are a priced commodity in Assam.

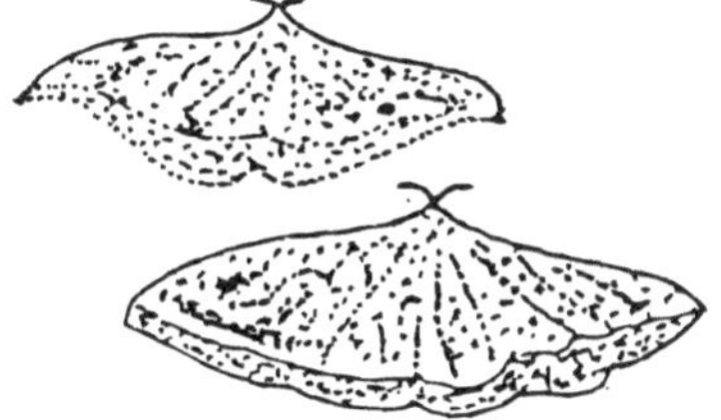

Fig. 3.12: Muga silk moth

The regional Muga Research station at Boko is developing appropriate technology to replace the traditional methods as this results in low productivity.

Rearing of Silkworms:

The female starts laying eggs soon after copulation and she is allowed to lay eggs on plain paper or mulberry leaves. At least one or two days before the hatching of eggs a small grid of wire mesh is placed above the trey of the eggs in such way that the newly emerged loopers when migrate should reach in this tray. The trays are removed at regular intervals and thus the larvae are shifted from the hatching age. They feed voraciously and grow rapidly. In this way they are fed till they pupate. Prepupal stage of larvae are again migratory so a close watch is needed to cheek them. Their date of silk formation is noted and the cocoons of same age are kept together when the cocoon is fully formed. Some of them are allowed to emerge as adult and continue the race but rest of them are killed. Killing is done by fumigation or exposed to sun for 2-3 days or boiled in hot water.

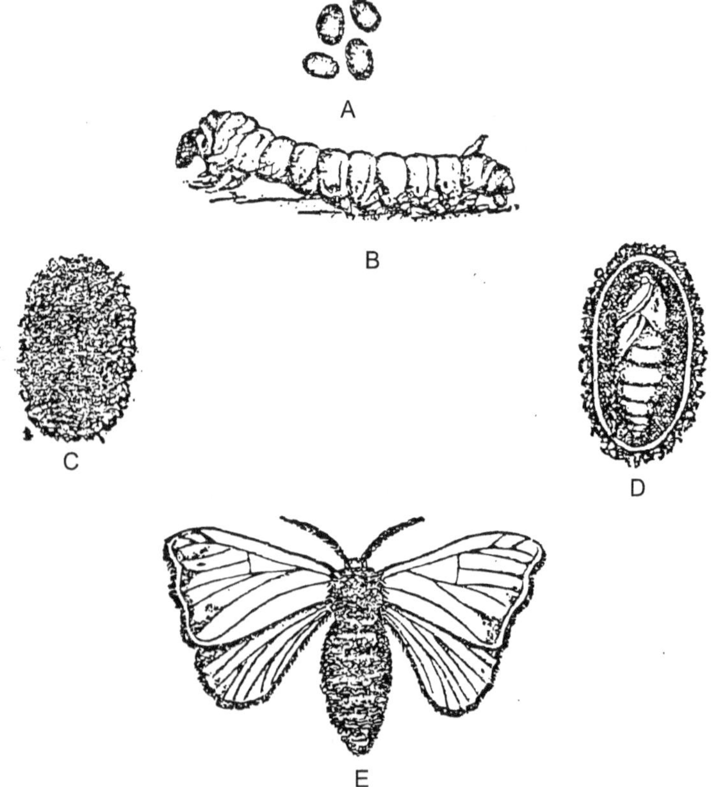

Fig. 3.13: Life stages of silkworm - Bombyx mori, Linn.
A - eggs; B - caterpillar; C - cocoon;
D - pupa insides a cocoon; E - adult.

But boiling of cocoons is beneficial as it soften, the silk which helps in reeling the thread as well as the pupae are killed soon without much problem. The outer layer of silk never comes in proper from so their cut pieces are collected and twisted together whereas the silk in inner layer is always in arranged and organised way and can be reeled easily. This is called raw silk. The raw silk is then taken for further processing and treating with chemicals for more strength and colouring. Sericulture is systematic rearing of silkworm and obtaining silk.

Lac Insect:

Lac is known to Indian people since time immemorial. It is reported in our vedic literatures as well as in Mogal period. It is supposed to be the native insect of India and today also the Indian states produce more than 85% of total lac produced in the world. Its production, management, distribution and culture was well known to Indian citizens in vedic period but today it is done in more scientific and organised way.

Lac is produced by an insect *Tacchardia lacca* which is small in size and colonial in habit. Lac is a resin like product which the lac insect produces and itself it is surrounded in it as a protective measure of the insect which harms the tree also.

Lac is produced from the posterior part of the female insect only but males do not produce lac. The females are larger than males with oval or pear shaped body enveloped in a cell like covering. It is almost flattened reduced legs, antennae, wings and other body appendages but piercing and sucking type of mouthparts are highly evolved and well developed. The females are bigger with a small pore at the hind end of abdomen.

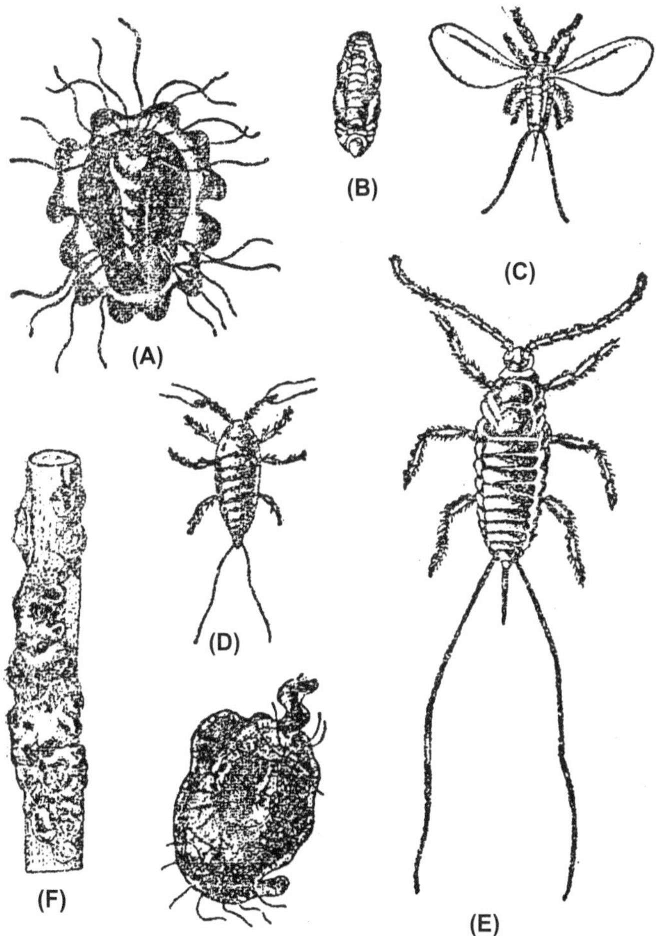

Fig. 3.14: Lac Insect

The males are small, cylindrical with a pair of weak wings, three pairs of legs, piercing and sucking types of mouthparts. The caudal style is present. The organs are fully developed and are not degenerated like the females.

The lac insects are generally found on the common trees like Ber, Pippal, Palas, Kusum, Babhool etc.

These insects are always associated with the new succulant branches of trees and suck the juice regularly. The twig harbouring these insects generally dries up and is defoliated. Climatic conditions and type of tree play a vital role in the production of lac and its quality.

Life Cycle: The female after producing the required amount of lac stops feeding and its body shrinks considerably. It lays about 300-400 small, rounded eggs at a time and the female die leaving an encrustation full of eggs. The hatching of eggs is governed by the temperature and humidity.

The larvae come out in large number from encrustation. They are equipped with piercing and sucking type of mouthparts. They move on succulent shoot where they can feed easily. The larvae settle in a close association with each other. Therefore twig is not seen.

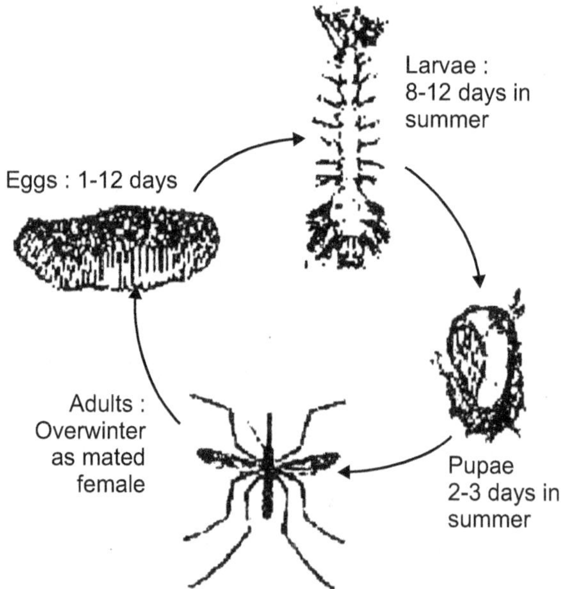

Fig. 3.15: Life Cycle of Lac Insect

They feed upon the sap and secrete pink coloured resin like substance from their dermal glands. This becomes hard when comes in contact with air and is known as lac. Lac deposition is done in all parts of the body except the mouthparts, spiracles and anal pores. This covering is called 'cell' which protects the larva and all the post embryonic stages are completed inside the cell. The larvae moult thrice inside the cell and become sexually mature. The male and female are almost alike but they become distinct within two weeks. The male is marked by its elongated cell. It contains two opening at anterior end. Female cell is rounded with irregular marine which with six holes through hair which like fibres may emerge.

The male cell produces winged and wingless males which are active and come out of the cell, walk over lac crustation of the female and fertilize the female. Both males are capable of fertilizing the female but the winged males move faster than the wingless males. They possess well developed legs, wings, and other appendages. Whereas the female larva after three moultings becomes sexually mature, but the appendages are reduced. The sexually mature female gets fertilized through the anal opening present in the cell. It grows very fast and produces more resin like substances, white, thin, fibrous, hair like processes come out of the thread and give a white appearance.

The females live longer than males but never come out of their cells. Depending upon the completion of life cycle stages and production of lac generally two crops are produced in a year. One is called *Baishakhi* and other *Kartiki* crop. The period of crop depends upon the climatic conditions, rainfall, temperature and humidity as well as the type of host plant available. Various type of plants have different period of defoliation and emergence of new leaves. So they directly govern the production of lac and maturation of post embryonic stages. Generally, eggs are laid in summer and winter seasons which get matured with stipulated period and produced in two seasons.

Adverse climatic factors like hot wind, high rainfall, drought, high temperature and humidity, and extreme cold, foggy weather are the enemies of lac insect which kill newly emerged males. Rats, squirrels, bats feed on lac insects and damage the deposited lac. Some moths and larvae also eat the eggs of lac insect.

The lac contains resin, dye and hard wax. A small amount of water and mineral is also found in it. The composition of lac is, Resin 60-85%, dye 2-10%, wax 6%, aluminous protein 5-8%, minerals 6-7% and water 3-4%.

Artificially lac production is obtained with more scientific way. Artificial inoculation of lac in the plant gives better and regular supply, as well as the amount of lac received is also much higher than that of natural inoculation.

Use of Lac: Lac is used in a variety of ways, bangles, toys, wood work, sealing waxes, bad conductors, records, varnishes, polish, inks, filling the spaces, and silvering the mirror are the some of the uses of lac.

3.5.2 Harmful Insects : Female Anopheles Mosquito, Red Cotton Bug, Rice Weevil

Female Anapheles Mosquito:

Systematic Position:

Kingdom:	Animalia	Order:	Diptera
Subkingdom:	Metazoa	Suborder:	Nematocera
Phylum:	Arthropoda	Super family:	Culicidea
Subphylum:	Manibulata	Family:	Culicidae
Class:	Insecta	Genus:	*Anopheles*
Subclass:	Pterygota	Species:	culcifacies

There are about 30 genera and 2700 species of mosquitoes. The most important mosquito genera are *Anopheles, Culex, Aedes, Mansonia* in India and they are related to disease transmission. Well known Indian species are *Culex pipens, Culex quainquefasciatus (= fatigans), Aedes aegypti, Aedes albopictus, Anopheles culicifacies* and *Anopheles stephensi, Mansonia annulifera* and *Mansonia uniformis*.

Mosquitoes are found all over the world. They only require water, plants and their host species. They occur in deep mines and at high altitudes. They develop in ice cold water and also found in hotsprings. Some of them require only clean water for breeding. Some mosquitoes prefer polluted and filthy water. Many species live in damp forests and marshes. While others are dwellers of crowded human habitations.

Malaria:

Malaria is transmitted by the bites of certain species of infected female Anapheline mosquitoes. A single female anopheles mosquito during her life time may infect several persons. It is infective when sporozoites are present in the salivary glands otherwise it is not infective.

Following are the conditions necessary for transmission of Malaria:

(1) A reservoir of infection, i.e. presence of individuals with a sufficient number of mature viable, male and female gametocytes in their blood.

(2) Presence of atleast one species of anopheline mosquito which is capable of transmitting malaria.

(3) Climatic conditions should be favourable for the development of sexual cycle in the insect vector and the vector should live long enough to allow the sexual cycle to be completed.

(4) The presence of susceptible human beings to whom the infection may be transmitted.

Apart from mosquito bite malaria may be transmitted by blood transfusion. Malaria may also be transmitted in drug addicts by shared syringes.

The length of time between the bite of an infected mosquito and the first attack of fever is usually not less than 10 days. This is called incubation period. This period varies according to the species of the parasite. Usually about 12 days are required for the development of *P. falciparam* infection, 14 to 15 days for *P. vivax* and *P. ovale* and upto one month for *P. malaria* infection with some strains of *P. vivax* the incubation period may be delayed for as long as 6-9 months.

There are three stages seen in Malaria.

(i) **Cold Stage:** Sudden onset of fever with rigor and sensation of extreme cold accompanied by shivering. The patient desires to cover himself with blankets.

(ii) **Hot Stage:** Patient feels burning hot and casts off his clothes. There is intense headache.

(iii) **Sweating Stage:** Fever comes down with profuse sweating, there is enlargement of spleen and secondary anaemia with a tendency to relapse.

Malaria is widespread in the tropical and subtropical regions. The WHO reports there were 198 million cases of malaria worldwide in 2013. This resulted in an estimated 5,84,000 to 8,55,000 deaths the majority (90%) of which occurred in Africa. In India also every year millions of people die due to malaria.

Red Cotton Bug:

Class	–	Insecta
Order	–	Hemiptera
Family	–	Pyrrhocoridae
Genus	–	*Dysdercus*
Species	–	*cingulatus = koenigii* (Fab)

The red cotton bug has wide distribution, it is a minor pest in cotton growing region of northern India particularly Punjab and Uttar Pradesh. This pest also occurs throughout the Maharashtra state but is minor importance. It is commonly known as a "*cotton stainer*".

Host Plants: Cotton, bhendi, ambadi, hollyhock and several other malvaceous plants.

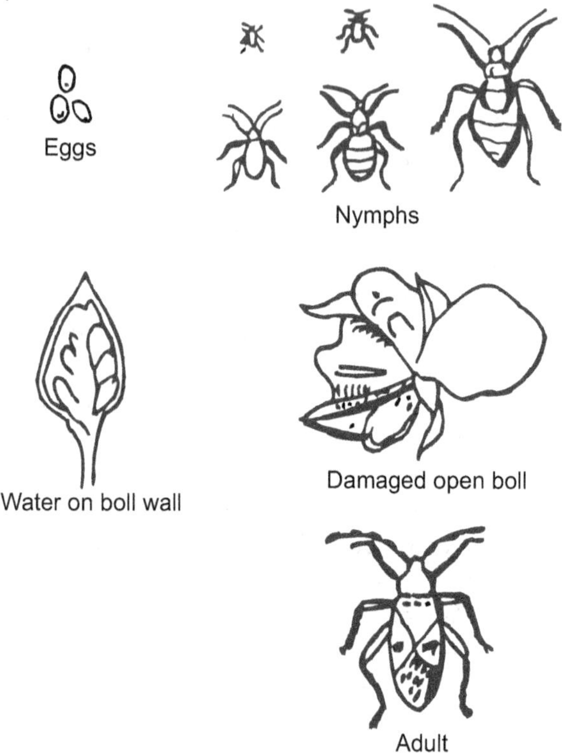

Fig. 3.16: Red Cotton Bug

Identification Marks:

The adult bug measures about 12-15 mm in length. The females are longer (15 mm) than the males (12 mm). It is blood red in colour except eyes, scutellum, anal, antennae which are black coloured. Besides, there is a black spot on each of the membranous forewings. A series of white transverse bands are present on the ventral side of the abdomen. Mouthparts are adapted for piercing and sucking. They form a straight beak or rostrum. The nymphs are smaller than adults and are wingless.

Life Cycle:

The mature female lays eggs during spring in clusters of 70-80 eggs each under the moist soil surface; fallen leaves and in crevices. The eggs are spherical, yellowish-white about 1.2 mm in length. After 7 days of incubation period and moist weather, eggs are hatched into active 1 mm long red coloured nymphs which are resemble the adult except size and absence of wings. The nymphs feed gregariously on the cotton bolls. The nymphs undergo 5-moults within 49-89 days to reach adult stage. In winter the life of the adult is about three months but in summer it is varied. Pest breeds on cotton from August-November; takes shelter under leaves or debris from December-middle of March and feeds on bhendi from April-July.

The life cycle of bug is completed within six to eight weeks.

Nature of Damage:

Both nymphs and adults, suck the cell sap from the leaves and tender shoots and impair the vitality of the plant. If the attack is severe, bolls open badly and the lint is of poor quality. In addition they also feed on the seeds and lower their oil content and low percentage of germination; such seeds are unfit for sowing. The lint is stained by the excreta of bugs or by their body juice as they are crushed in the ginning factories.

Control Measures:

1. Cotton field should be ploughed to expose eggs to sunlight.
2. Insects should be hand picked and killed in kerosinised water.
3. The crops of bhendi should be sown as trap crop and pests collected there, should be destroyed.
4. Moistened cotton seeds should be hunged up at different places in the field where bugs congregate, they may get killed in the kerosene mix water.
5. Spraying of Malathion 0.05% is effective to control the pest.
6. Spraying of 1 litre endosulphan 35% EC, 0.25 litre phosphamidon = 100% EC or 1 litre Fenitrothion 100% EC per hectare is very effective or reduces pest population.

Rice Weevil:

Pest of Stored Grains:

The storage of food grains has been a long practice with cultivators and traders. Considerable losses both in quality and quantity of food grains take place in storage due to number of factors. Organisms like insects mites, rodents, fungi, and bacteria are directly responsible for causing loss in stored products. Insects and mites are most hazardous to stored grains because they are general feeders but some insects prefer certain grains.

It is estimated that about 7-10% store grains are lost every year due to insect damage in India. The annual losses caused by insect pests destroying our stored and packaged food run into several hundred crores of rupees.

Much of the damages to store grains by insect pests is done directly to the kernels. The degree of damage depends on three factors : (i) *Moisture* content of the stored grains. (ii) *Temperature* inside the storage place and (iii) *Oxygen level.* Following are examples of major insects pests of stored grains.

Sitophilus Oryzae Linn:

Common name	–	Rice weevil
Class	–	Insecta
Order	–	Coleoptera
Family	–	Curculionidae
Genus	–	Sitophilus
Species	–	Oryzae

This is a very serious pest of stored grains as well as grains in farm storage. It is cosmopolitan in distribution.

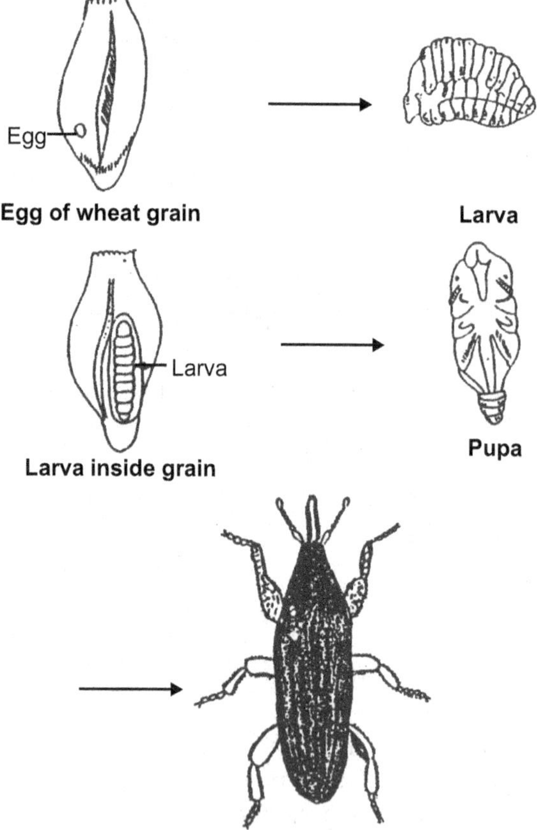

Fig. 3.17: Life Stages of *Sitophilus Oryzae* (Gram)

Identification Marks:

The adults are small weevil, about 4 mm in length with reddish brown, dark brown or almost black in colour with cylindrical body and a long curved rostrum. The head is prolonged into a slender snout with chewing mouth parts located at its tip. The functional wings hidden beneath the dark brown elytra with four light reddish or yellowish spots and the insect is able to fly. The life span of adult weevil is 4-5 months. The grubs are whitish in colour, small (i.e. 5 mm in length) and legless with yellow brown head. They are always found inside the kernels.

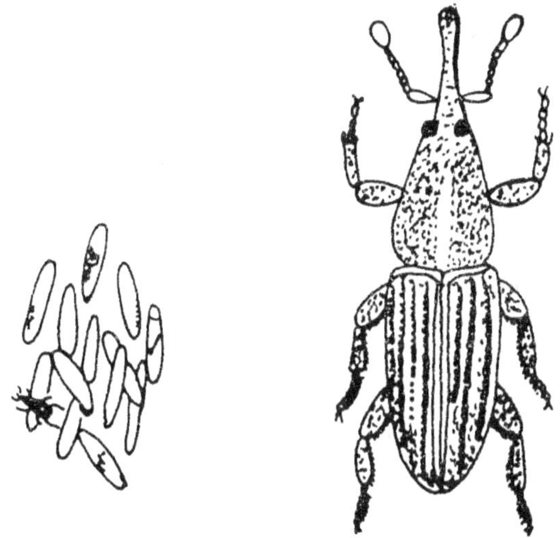

Fig. 3.18: *Silophilus Oryzae* **(Rice)**

Host:

S. oryzae prefer rice, they can be found feeding on wheat, maize and other grains.

Nature of Damage:

Generally infestation starts in grains only during storage. The *S. oryzae* feeds on whole grains of rice, wheat, jowar, bajri, maize, barley etc. Both adults and larvae feed voraciously on the grains so that the grain becomes unfit for consumption and seed purposes. A thin tunnel (hole) is formed by the grub from the surface towards inside of the grain. Circular exit holes and stained spots on the surface of the grain kernel is the symptom of damage. In case of heavy infestation, the grains become a mass of broken vegetable matter. The adults eat a small amount of grain, making shallow holes with rugged edges but the amount of damage thus caused is negligible as compared to the complete hollowing of grain by the larval stages results into hollowed grains leading to reducing in weight and food value.

Life Cycle:

The female starts laying eggs 5 days after the emergence. The female bores a small, round hole in the soft part of a grain by means of mouth parts and there she lays a single egg and sealing it with a mucilaginous secretion. The eggs are translucent, white, oval, minute (0.7 × 0.3 mm in size). A female lays a total of 300-400 eggs during her life-time. The incubation period varies from 4-7 days in summer to 6-9 days in winter.

The grub (larva) hatches out of the egg is legless (Apodous), translucent white with fleshy body and yellow-brown head. The young tiny grub bores into the grain kernel and lives within grain; feeding on its starchy content and hollows it out, leaving only the outer shell intact. There are four larval instars and the larval period lasts for 25-35 days. The full-grown grub makes a pupal cell inside the grain. The pupa is curved or humpy in appearance and takes 3-6 days to emerge as an adult. The emerged weevil immediately starts breeding and gives rise to new generation for destruction of grain. Several generations are possible during a normal storage season of grains. The adults live on an average for 2 to 5 months.

Control Measures:

To control the stored grain pest. Both the control methods are adopted i.e. the preventive and curative.

1. Preventive Method:

This control measure is employed to protect fresh stocks from the attack of pests.

(i) After harvesting, dry the grains in sunlight sufficiently so that the moisture content is reduced to less than 8% moisture because most of the stored grain pests cannot multiply in grain at such percentage of moisture.

(ii) After drying in sunlight, the grains store in new gunny bags so that grains will be free from infection.

(iii) Use of neem leaves, mercury, mixing of ash, powder of sweet flag rhizome (*Acorus Calamus* Linn), etc. in grains or smearing the grains with plant oils are the indigenous practices used by rural peoples for protecting the stored grains from the attack of different pests.

If the old bags are to be used, they should be fumigated with EDBR (Ethylene di-bromide) at 3 kg/100 cu.m for 5 to 7 days which helps to check the cross infestation. It also achieved by applying 5% BHC or 0.06% pyrethrum dust at 25 gm/sq.m or sprays of BHC, pyrethrum or malathion. The bags should then be stored in an insect free godown.

(iv) To make the insect free godown one or more of the following methods may be used.

(a) All dirt, refuse material should be removed and destroyed and all the cracks, crevices, holes in walls, floors or ceiling of the godown should be filled in with cement.

(b) The rat holes should be closed by filling them with cement or sand mixed with glass pieces.

(c) Fumigate the godown with EDBR.

(d) Spray with insecticides like Pyrethrum, Malathion or BHC to avoid the chance of contamination.

2. Curative Measures:

This measure becomes essential when the grains get infested with the stored grain pests.

(i) Sieving and cleaning removes all the stages of pests.

(ii) After above method the grains must be followed by sunning and fumigation with suitable fumigants.

(iii) The larvae and adults are killed by exposing them for 48 hours to the vapours of ethylene dichloride - carbon tetrachloride mixture under gas proof covers. Fumigation of infested grain with methyl bromide is also effective and kills all stages of the pest including eggs.

POINTS TO REMEMBER

- Arthropods are the animals with jointed legs.
- Body covered by chitinous exoskeleton.
- Open type of circulatory system.
- Gills, trachea, skin and book lungs are the respiratory organs.

- Excretion occurs by malpighian tubules.
- Subphylum mandibulata has three classes namely Crustacea, Arachnida, Myriapoda and Insecta.
- Arthropods are economically important animals.
- Honey bees, lac insects and silkworms are useful insects for man.
- Female anopheles mosquito, red cotton bug and rice weevil are harmful insects to man and stored grains.
- Biting and chewing, chewing and lapping, piercing and sucking, sponging and lapping and siphoning mouth parts are present in insects.

EXERCISES

1. Give an account of general characters of phylum Arthropoda. Give an outline of classification of Arthropoda with characters and suitable examples of each class.
2. Give distinguishing characters of the following classes with suitable example.
   (a) Crustacea, (b) Arachnnida, (c) Myriapoda, (d) Insecta,
3. Give an account of different mouth parts in insects.
4. Write an essay of economic importance of Arthropods.
5. Give an account of beneficial insects to man.
6. Describe different harmful insects to man.
7. Write short notes on :
   (a) Biting and chewing types of mouthparts
   (b) Chewing and lapping types of mouthparts
   (c) Piercing and sucking types of mouthparts
   (d) Sponging and lapping types of mouthparts
   (e) Siphoning type of mouthparts
   (f) Economic importance of Arthropoda
   (g) Useful insects
   (h) Harmful insects
   (i) Silkworm
   (j) Lac insect
   (k) Honey bees
   (l) Anopheles mosquitoes
   (m) Red cotton bug
   (n) Rice Weevil.

✳✳✳

4

CHAPTER

PHYLUM - MOLLUSCA

CONTENTS

4.1 INTRODUCTION TO PHYLUM - MOLLUSCA

Mollusca means soft bodied animals. (Latin, molluscs-soft). Aristotle first used this term for cuttle-fish. In Latin soft nut enclosed in a thin shell is called mollusca. Thus, it is referring to the bivalve shell and soft bodied animal enclosed in the shell.

4.2 SALIENT FEATURES OF PHYLUM - MOLLUSCA

1. Molluscs are aquatic, mostly marine, some are freshwater and some terrestrial.
2. The symmetry is bilateral, however, gastropods and cephalopods lose their bilateral symmetry and become asymmetrical due to torsion or spiral twisting.
3. Body of the molluscs is soft, differentiated into four parts anterior head, dorsal visceral mass, ventral foot and mantle.
4. Epidermis is single layered, generally ciliated with mucous glands.
5. Muscular foot is present on ventral side which is locomotory organ and modified for creeping, swimming and burrowing.
6. A thin, muscular, fleshy fold covers the dorsal body wall called mantle or pallium. The space enclosed by mantle is called mantle cavity.
7. Shell is secreted by outer surface of mantle. Shell is hard, calcarious and it may be bivalved, univalved, spiral or cone like, internal or external, reduced or even absent in some animals.
8. Respiration by gills called ctenidia. Body surface, mantle or lungs are respiratory organs in terrestrial forms.

(4.1)

9.   Digestive system is complete. Buccal cavity contains a grasping organ, the radula with transverse rows of teeth.
10. Circulatory system is of open type.
11. Excretion is brought about by one or two pairs of sac-like kidneys.
12. Nervous system consists of paired ganglia, cerebral, pleural, pedal and visceral ganglia inter connected by commissures and connectives.
13. Tentacles, eyes, statocysts and osphradia are the sense organs.
14. Sexes are usually separate, some are hermaphrodite, fertilization is external or internal.
15. Segmentation (cleavage) is spiral. Development may include glochidium or veliger larva.

4.3 CLASSIFICATION OF PHYLUM - MOLLUSCA

The phylum mollusca is divided into six classes: Monoplacophora, Amphineura, Gastropoda, Scaphopoda, Pelecypoda and Cephalopoda.

Class-Monoplacophora:
(1)   Marine and they are called living fossils.
(2)   Body has bilateral symmetry and is covered by a spoon or cup shaped shell.
(3)   Head bears tentacles.
(4)   The disc like foot has a flat creeping sole.
(5)   The visceral mass is divided into five segments each with a pair of shell muscles, gills, aurides, nephridia and gonads.
(6)   Buccal cavity contains radula.
(7)   Stomach contains a crystalline style.
(8)   Nervous system lacks ganglia.
(9)   Sexes are separate.
Example: *Neopilina*.

Class-Amphineura:
(1)   They are marine, bottom dwelling primitive molluscs.
(2)   Body is bilaterally symmetrical, vermiform, elongated and flattened.
(3)   Head is not distinct, without tentacles and eyes.
(4)   Foot is large, ventral, flattened and useful for creeping.
(5)   Dorsal side of the body shows spicules or calcarious plates.
(6)   Respiration by gills.
(7)   Nervous system is primitive.
(8)   Excretion by pair of kidneys.
(9)   Sexes are separate or united.
(10) Fertilization is external and development includes a pelagic trochophore larva.
Examples: *Chiton, Neomenia, Chaetoderma*.

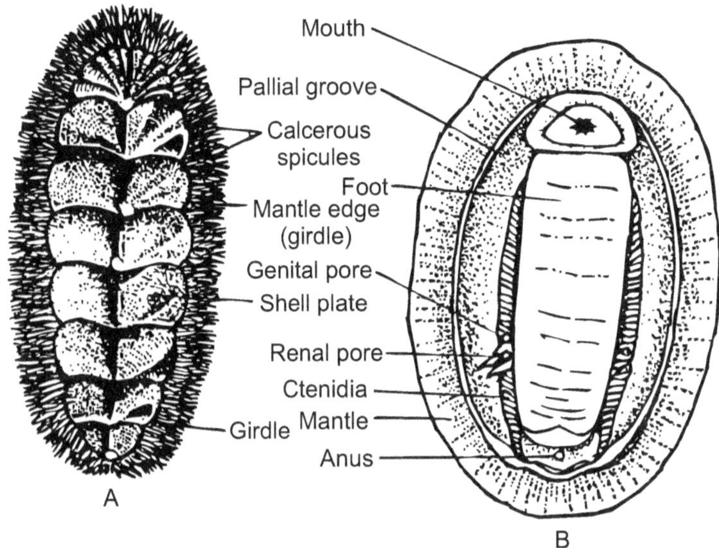

Fig. 4.1: Chiton

Class-Scaphopoda:

(1) Body elongated, worm like and bilaterally symmetrical

(2) Shell is tubular, curved and open at both the ends.

(3) Foot is conical and useful for digging.

(4) Head rudimentary and without eyes and true tentacles.

(5) Long, thread-like, prehensile, knobbed processes called captacula useful for capturing the food.

(6) Ctenidia or gills are absent.

(7) Respiration by mantle.

(8) Sexes are separate and life history includes a veliger larva.

Examples: *Dentalium* (Tusk shell) and *Siphonodentalium*.

Class-Gastropoda:

(1) Gastropoda includes marine, freshwater as well as terrestrial forms.

(2) Body is asymmetrical due to torsion.

(3) Head is distinct and bears tentacles and eyes.

(4) Foot is ventral, large and flat useful for locomotion and attachment. It bears operculum for closing the shell aperture.

(5) Shell is univalved and spiral, secreted by mantle.

(6) Buccal cavity contains odontophore with radula having transverse rows of chitinous teeth.

(7) Respiration by body surface or gills or lungs.

(8) Circulatory system is open and lacunar, heart has one or two auricles and one ventricle.

(9)   The excretory system consists of a single kidney.

(10) Sexes are separate or united. Development may be direct or includes trochophore and veliger larval stages.

Examples: *Pila, Haliotis, Patella, Cypraea, Helix, Lymnea.*

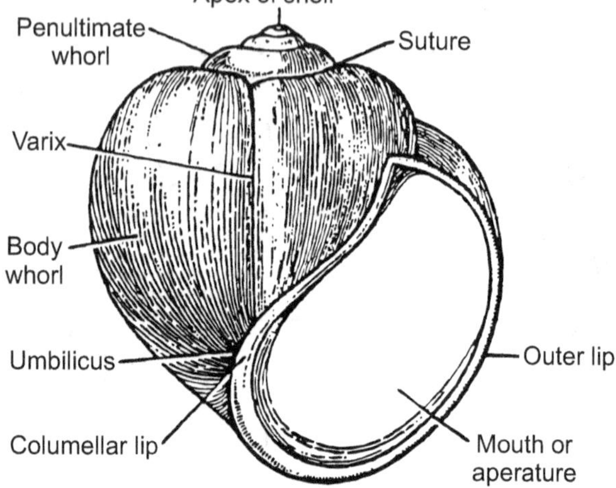

Fig. 4.2: *Pila globosa*

Class-Pelecypoda:

(1)   Majority of animals are marine, some are freshwater and generally lead sedentary or burrowing life.

(2)   Body is bilaterally symmetrical and laterally compressed.

(3)   Head is reduced without eyes and tentacles but bears a pair of labial palps.

(4)   Foot is ventral, large, muscular wedge shaped adapted for burrowing in sand and mud.

(5)   Shell consists of two (right and left) valves movably hinged dorsally and closed by one or two adductor muscles.

(6)   Mantle comprises two (left and right) lobes which are united dorsally but hanging free ventrally and enclosing spacious mantle cavity.

(7)   Inhalent and exhalent siphons are present for the entry and exit of water into and from the mantle cavity.

(8)   Respiration by ctenidia.

(9)   Alimentary canal is coiled tube without radula.

(10) Circulatory system consists of sinuses and vessels, heart has two auricles and one ventricle enclosed in a pericardium.

(11) Excretion by paired kidneys.

(12) Sexes are separate. Development includes trochophore, veliger and glochidium larvae.

Examples: *Unio, Mytilus, Pecten, Pinctada (Oyster), Solen.*

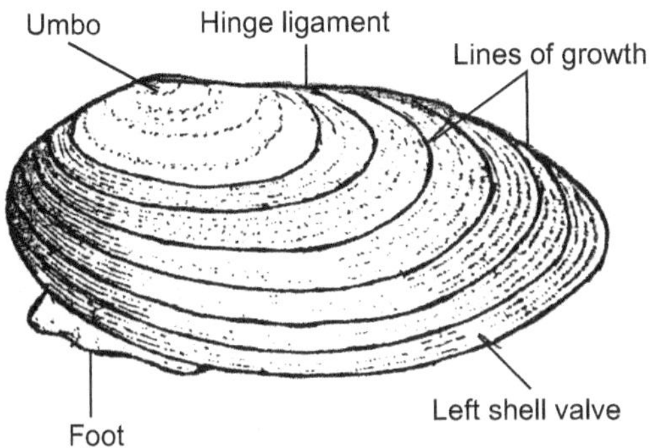

Fig. 4.3: *Unio*

Class-Cephalopoda:

(1) They are marine, fast swimming, highly organised, predaceous animals.

(2) Bilaterally symmetrical body having head and trunk.

(3) Head is prominent and bears pair of eyes and mouth.

(4) Foot is partly modified into numerous sucker bearing arms or tentacles surrounding the mouth (hence cephalopoda). The tentacles may be 8 or 10.

(5) The trunk is uncoiled.

(6) Shell may be external and well-developed or internal and reduced or absent altogether.

(7) Mouth is provided with horny jaws and radula.

(8) Respiration by two pairs of ctenidia.

(9) Excretion by two pairs of kidneys.

(10) Sexes are separate.

(11) In male, one arm is hectocotylized. Serves as a copulatory organ.

(12) Development is direct.

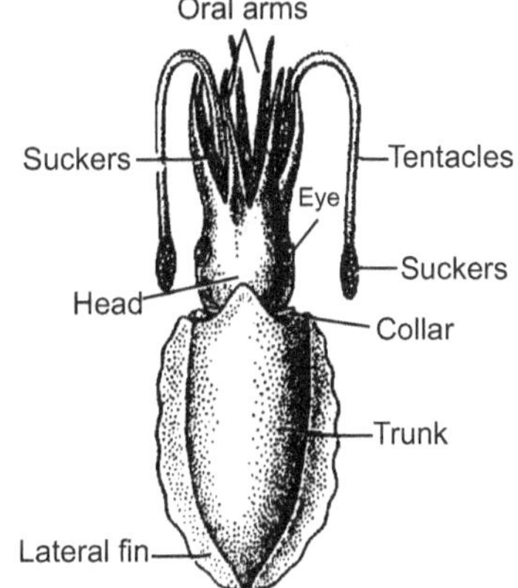

Fig. 4.4: Sepia

Examples: *Sepia, Loligo, Octopus, Nautilus.*

4.4 ECONOMIC IMPORTANCE OF MOLLUSCA

There are significant number of molluscs animals in invertebrate fauna of world. Most of the molluses are of major interest to man and about one lac species of molluses are of economic importance. There are many reasons of interest of man in molluscs. The great variety of shells, their colouration, symmetry, shapes and sizes have attracted millions of collectors, young and old. These shells are used in ornaments, as well as in making fancy gift articles. These attractive shells are also used as show pieces for decoration. The shells are also used for many other purposes. The inner side of the clam shells are with iridescence and hence from such shells buttons are made. These shells are also used in making knife handles and many other ornamental objects.

Some molluscs like pearl oysters produce natural pearls; which are highly precious having values ranging up to several hundred rupees each. Some rare and colour pearls are very costly. Man has now employed these molluscs for generating cultured pearls and by selling pearls making profit. In pearl industry Japan is now at the top in the world.

All the molluscs are not beneficial to man. Some mollucs are beneficial to man. Some molluscs are harmful to man as pest and do untold damage every year. Many snails and slugs feed on cultivated plants. Some species are injurious to crops, as well as human beings. Certain freshwater species of snails serve as an intermediate hosts for parasitic trematode worms (e.g. liver fluke) of man and his domestic animals. The molluscs like teredos can cause serious damage to wooden ships and wharves. They make their shelter in wooden material.

The molluscs shells are useful for detecting geological ages of various stratifications or layers of earth. The plaeontologists use these shells for the study of chronological time clocks for judging the geological ages of earth layers. The hard molluscs shells are preserved as the best fossil material. From the study of these fossils the phylum mollusca is quite ancient with very complete fossil record dating back to 550,000,000 years. Therefore, mollusc fossils are most important and indispensable to the branch of science Geology in general and to the science of Petroleum Geology in particular. Through the detailed study of marine molluscs fossils, the petroleum geologists can easily locate the valuable petroleum deposits under sea. Petroleum deposits are often lie several kilometres beneath the surface of sea. In Maharashtra, Bombay high is the best example of oil deposit under the bottom of sea.

Pearl Industry:

Pearl oysters are economically very important bivalves because they have ability to produce high quality pearls as gems. pearl oyesters belong to the genus *Pinctada* under family Pteriidae of class Bivalvia. Following are the some of the important species found in Indian waters.

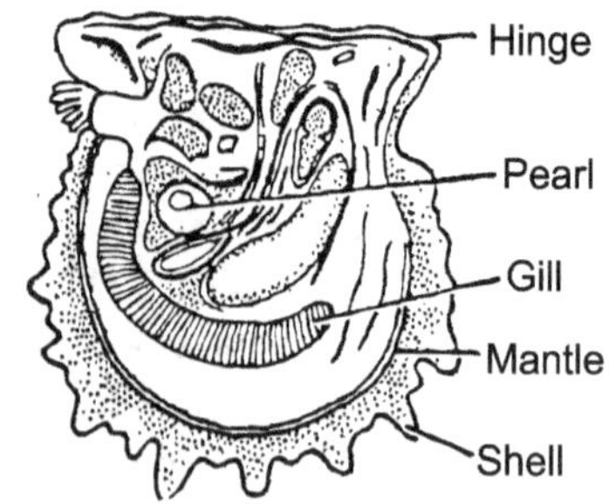

Fig. 4.5: Pearl formation

1. *Pinctada vulgaris*
2. *Pinctada fucata*
3. *Pinctada chemitzii*
4. *Pinctada sugillata*

P. vulgaris is common oyster distributed in Gulf of Kutch, Gulf of Manmar and the Pak bay. Some are distributed Persian Gulf, Red Sea, *Sri Lanka.*

The pearl oyester forms 'pearl banks' along the Indian coasts extend from Cape Comorin to Rameshwarm Island with the most productive ones near Tuticorin. They are at 10-12 fathoms of water depth and at a distance about 20 km from shore. They produce millions of oyesters worth several lakhs of rupees annually. The oyester in the Gulf of Kutch are found on reef near Jamnagar and easily collected by fishermen.

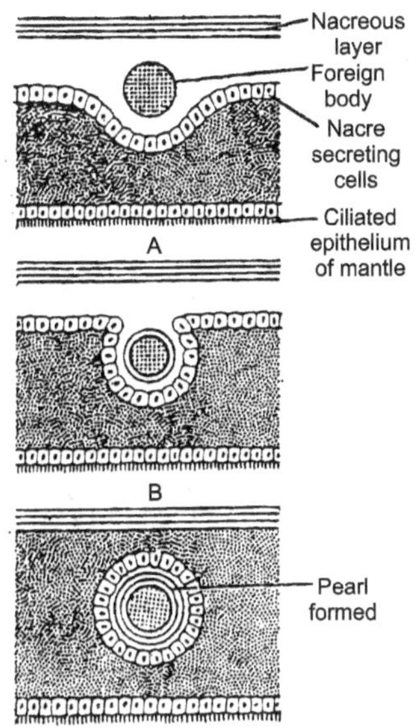

Fig. 4.6: Stages in pearl formation :
A - Primary stage;
B. - Progressive stage; C - Final stage.

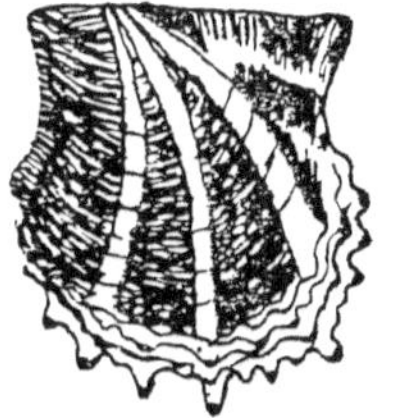

Pinctada margaritifera   *Pinctada chemnitizii*   *Pinctada fucata*

Fig. 4.7: Pearl Oysters

Pearl oysters are attached to the substratum by shell with a strong byssus. They are filter feeder like other bivalves feeding on diatoms drawn in with the ciliary current. It can grow upto 65 mm in about 5.5 years and breeds twice a year, viz. April-May and September-October. Sexes are separate and fertilization is external and larval period is of 3 weeks.

Pearl Formation: When any foreign particle, parasite, sand grain enters the body and adheres mantle cavity, the mantle epithelium starts to secret and concentric layers of nacre around as a defensive mechanism and the particle is completely enclosed gradually. Such several layers are formed and pearl is formed. This is natural phenomenon but now purposely sand grains are injected in mantle wall and animal forced to form pearl is called cultured pearl. By this method pearls are obtained on large scale and hence pearl industries are formed in many countries. Japan is ranking first in pearl industry. For this pearl oysters are collected and reared in cages. Then foreign particles are inserted in the mantle and again cages are suspended in water and pearls are cultured in the body of oysters.

In natural conditions the oysters are collected by divers and are auctioned in lots of 1000 oyesters each. Then they are allowed to rot for few days, then opened and washed to collect the pearls contained in them. There are now many pearl fisheries on the coasts of India. Naturally, occurring pearls may be small or large, smooth and of fine luster, fetching high prices. Induced pearl formation is profitable. Japan, Australia, Philippines and many other countries including India are using these techniques. In CMFF at its oyester farm training of pearl culture techniques is given to scientists and technicians.

The pearl is a concretion of $CaCO_3$ in an organic matrix like the macreous layer secreted by the mantle on the inner surface of shell valve.

There are many other species which produce pearls include window pane oyster *Placenta placenta*, Ear-shell *Haliotis*, sea muscle *Mytilus*, *Placenta margaritifera*, *P. maxima* are giant species are found in muddy bays in Kutch, near Mumbai, near Kakinada in Andhra Pradesh. The best quality pearl is called *Lingha* pearl. Pearls obtained from freshwater bivalve are not valuable as marine oyster pearl.

The bivalve shells having brilliant silvery sheen called mother of Pearl are also collected from various parts of world for manufacture of buttons and other fancy articles.

POINTS TO REMEMBER

- Molluscs are soft bodied animals.
- They are aquatic and terrestrial.
- Respiratory organs are gills or ctenidia.
- Circulatory system is open type.
- Sexes are separate but some are hermaphrodite.
- This phylum is classified into six classes.
- Molluscs are economically important animals.
- Some molluscs are beneficial and some are harmful.
- Pearls are formed by pearl oysters.
- Japan is at the top in pearl production.
- Pearl is concentration of $CaCO_3$.
- There are many species of oysters which produce pearls.

EXERCISE

1. Give an account of general characters of Phylum Mollusca.
2. Give an account of classification of Mollusca.
3. Give an account of economic importance of Molluscs.
4. Describe the process of pearl formation in Pearl oyster.
5. Write short notes on:
   (i)   Give the characters and examples of Class Monoplacophora.
   (ii) Give the characters of Amphineura.
   (iii) Give the characters of Gastropoda.
   (iv) Give the characters of Pelecypoda.
   (v)   Give the characters of Cephalopoda.
   (vi) Economic importance of Molluscs.
   (vii) Pearl formation.

5
CHAPTER

PHYLUM - ECHINODERMATA

CONTENTS

5.1 INTRODUCTION TO PHYLUM - ECHINODERMATA

The name Echinodermata literally means 'spiny skinned animals' (Greek; echinus-spiny, derma-skin). The skin or test of these animals bears prominent spines.

5.2 SALIENT FEATURES OF ECHINODERMATA

1. Echinoderms are exclusively marine.
2. They are gregarious, mostly free-living, slow-moving (creeping), some are pelagic and few are fixed.
3. Animals show pentamerous radial symmetry and larvae show biradial symmetry.
4. Body shape may be star-like, globular, spherical, discoidal or elongated, flattened with oral and aboral surface.
5. Ambulacral grooves are present on the body surface.
6. Endoskeleton consists of closely fitted plates forming a shell or theca or test. These plates are hard and calcarious.
7. Exoskeleton is made of movable calcarious spines.
8. Body cavity or coelom is large, lined by ciliated peritoneum.
9. Presence of water vascular or ambulacral system is the characteristic feature of echinoderms. Sieve plate like madreporite is present which is hydropore.

10. Tube feet or podia are useful for locomotion, food capture and respiration.
11. Respiration occurs through variety of structures such as papulae (skin gills), peristomial gills, genital bursae and cloacal respiratory tree.
12. Pedicellariae protect the delicate papulae or skin gills.
13. Circulatory (haemal) system is reduced and lacunar. Heart is absent.
14. Excretory system is absent.
15. Nervous system is primitive, lacks a brain, but consists of a circum oral ring and radial nerves.
16. Sense organs are poorly developed and consist of statocysts, pigment eye-spots and tactile tentacles.
17. Sexes are separate, no sexual dimorphism.
18. Reproduction is entirely sexual, fertilization is external.
19. Development includes microscopic ciliated, transparent, free swimming larval stage.
20. Some echinoderms reproduce asexually by transverse fission and some show autotomy by which lost body part can be regenerated.

5.3 CLASSIFICATION OF ECHINODERMATA

Phylum: Echinodermata is divided into two subphyla: **Eleutherozoa** and **Pelmatozoa**.

Subphylum-Eleutherozoa:
(1) They are free living without stem or stalk.
(2) The mouth lies on the surface facing downward and anus on aboral surface.
(3) Body is pentamerous.
(4) Tubefeet or podia are locomotory as well as food gathering organs.
(5) Nervous system is located on oral surface.

This subphylum is divided into four classes: Asteroidea, Ophiuroidea, Echinoidea and Holothuroidea.

Class-Asteroidea:
(1) Free living, slow-creeping and predaceous animals.
(2) The body is flattened, star shaped or pentagonal, radially symmetrical and differentiated into a central disc and arms.
(3) The arms are usually five in number but their number may vary up to 50.

(4)　Oral and aboral surfaces are distinct.

(5)　Endoskeleton consists of separate ossicles.

(6)　The oral surface bears the mouth and five narrow open ambulacral grooves.

(7)　Two to four rows of locomotory tube feet or podia are present in each ambulacral groove which are retractile and provided with terminal suckers.

(8)　Respiration by papulae.

(9)　Movable pincer-like spines called pedicellariae are present.

(10) Sexes are separate. Development includes either bipinnaria or brachiolaria larva.

Examples: *Asterias* (Sea-star), ***Pentaceros, Astropecten.***

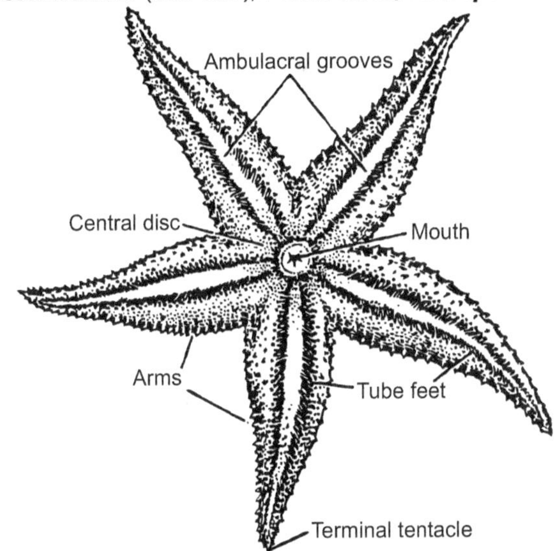

Fig. 5.1: Asterias or Sea star

Class-Ophiuroidea:

(1)　Body is flattened and star shaped.

(2)　Arms are usually five, long slender, cylindrical, joined and highly flexible.

(3)　Ambulacral grooves are absent.

(4)　Pedicellariae, skin gills and special sense organs are absent.

(5)　Madreporite is present on oral surface.

(6)　Tubefeet are not locomotory but useful for respiration and touch.

(7)　Locomotion is effected by slender arms.

(8)　Sexes are separate. Development includes free-swimming larva pluteus.

Examples: *Ophiothrix* (Brittle star) *Astrophyton, Ophioderms, Goragonocephalus.*

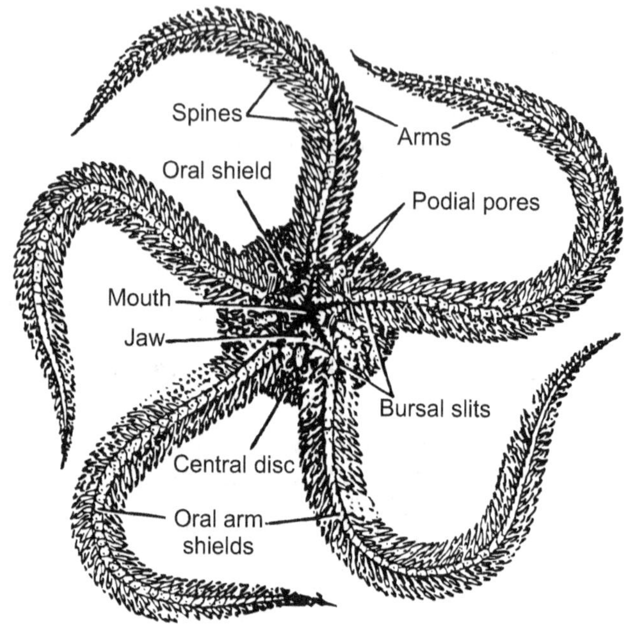

Fig. 5.2: Brittle star

Class-Echinoidea:

(1) The body is globular, heart-shaped or disc like, without arms.

(2) Endoskeleton forms a shell or corona of close fitting, immovable calcarious ossicles.

(3) The outer surface of shell is covered by long, movable spines and are used in locomotion.

(4) Outer surface shows alternative 5 ambulacral and 5 interambulacral zones or areas.

(5) Oral and aboral sides are distinct, mouth is on oral side and aboral surface carries the anus and madreporite.

(6) Ambulacral grooves are absent.

(7) Tubefeet protrude in two rows on each ambulacral zones having ampullae and suckers and are used as locomotory, respiratory and tactile organs.

(8) Pedicellariae are stalked and three jawed.

(9) Mouth has five teeth with elaborate system of ossicles, all forming the characteristic Aristotle's lantern.

(10) Sexes are separate. Development includes free swimming larva.

Examples: *Echinus* **(Sea urchin),** *Clypeaster* **(Cake-urchin),** *Echinocardium* **(Heart-urchin).**

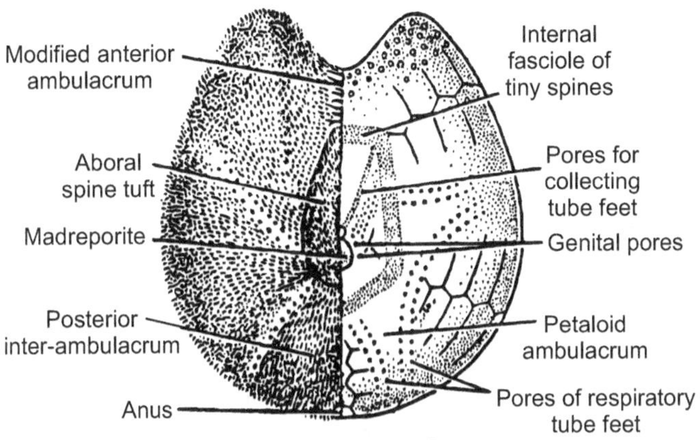

Fig. 5.3 Echinocardium

Class-Holothuroidea:

(1) Body is elongated, cylindrical and without arms.

(2) Endoskeleton is reduced to microscopic ossicles or spicules.

(3) Skin is soft, thin or leathery without spines and pedicellariae.

(4) Oral and aboral ends are distinct. Oral end is anterior and has the mouth surrounded by a ring of refractile, sometimes branched tentacles. These are modified tube feet and are often called buccal poidia.

(5) Aboral end is posterior and has the anus.

(6) Locomotory tubefeet usually present, occupying five ambulacral areas. Ambulacral grooves are absent.

(7) Alimentary canal is long coiled and cloaca usually having respiratory tree.

(8) Sexes are separate and development includes auricularia larva.

Examples: *Holothuria* **(Sea-cucumber),** *Thyone, Synapta, Stichopus.*

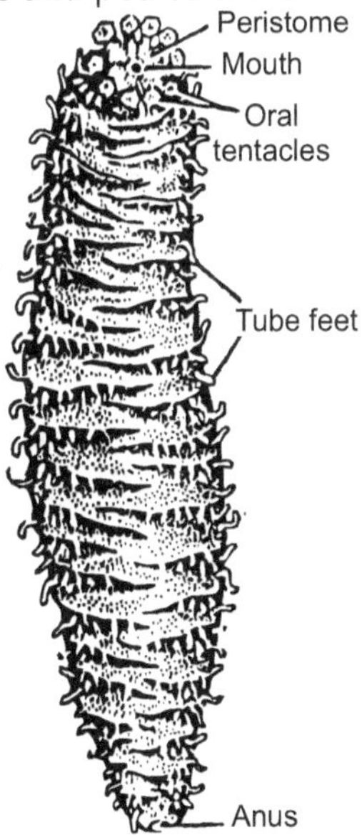

Fig. 5.4 Sea cucumber

Subphylum-Pelmatozoa:

(1) Mostly extinct animals.

(2) In early life, they are attached by aboral stalk supported by rows of calcarious ossicles.

(3) Oral surface is directed upwards and bears both mouth and anus.

(4)   Tubefeet are ciliated, food catching and without suckers. They function only as respiratory and tactile organs.

(5)   Spines, madreporite and pedicellariae are absent.

(6)   Main nervous system is aboral.

This subphylum includes a single living class namely, **Crinoidea.**

Class-Crinodea:

(1)   Extinct or living forms, usually attached permanently or temporarily by a jointed stalk to the sea bottom. Many are free swimming and without stalk.

(2)   Body is enclosed in a cup like theca, pentamerous with upwardly directed oral surface and downward aboral surface.

(3)   The arms are long, slender usually five or ten in number, branched with small alternating branches, the pinnules.

(4)   Ambulacral grooves are ciliated radiate out from mouth on oral side of arms and pinnules to their tips.

(5)   Calcarious endoskeleton is present.

(6)   Tubefeet are without ampullae and suckers and are useful for excretory, respiratory and tactile function.

(7)   Pedicellariae, spines and madreporite are absent.

(8)   Sexes are separate, development includes doliolaria larva.

(9)   Great power of regeneration is exhibited by these animals.

Examples: *Antedon* (Feather-star), *Sea-lily* (Metacrinus)

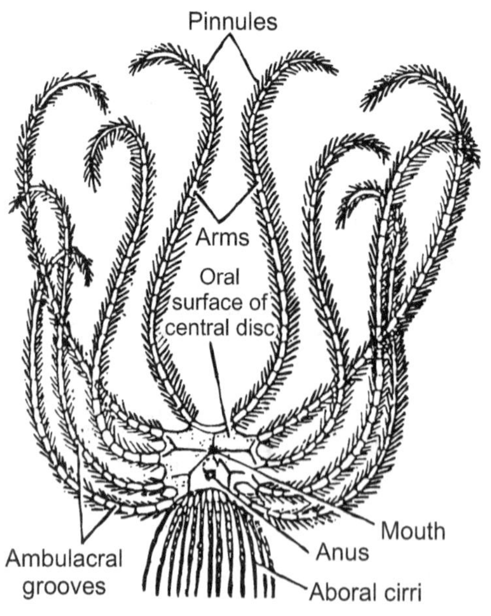

Fig. 5.5 Antedon

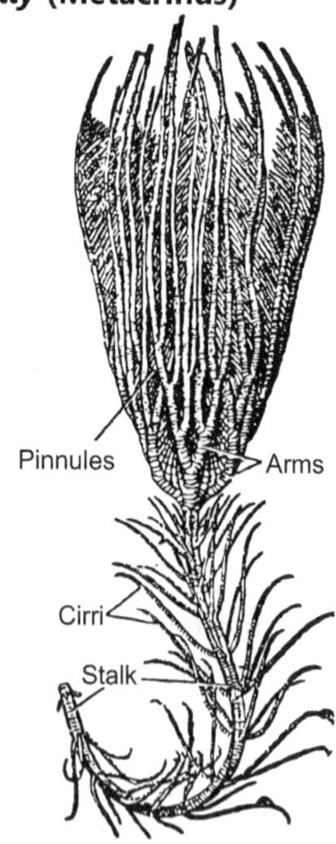

Fig. 5.6 Metacrinus

5.4 TYPE STUDY : *Asterias rubens (Sea Star)*

Asterias or Starfish is the most commonly known and widely distributed echinoderm. However, the common name starfish is somewhat misleading as it neither resembles a fish nor its shape is like a true star. Therefore, some authors renamed it as sea-star because it is star like in form and occurs in the sea.

There are little over one thousand species of starfishes which differ in certain details but as a class are more similar in structure and habits. The genus *Asterias* comprises nearly 150 species of which some important species such as *A. rubens* found on the English and North European coasts. *A. vulgaris* on the North Atlantic coast of North America; *A. forbesi* on the eastern sea shore from Maine to the Gulf of Mexico and *A. amurensis* found in the Behring sea, Japan and Korea.

Systematic Position:

Phylum	:	Echinodermata
Sub-phylum	:	Eleutherozoa
Class	:	Asteroidea
Order	:	Forcipulata
Family	:	Asteriidae
Genus	:	*Asterias*
Species	:	*rubens*

Habitat: *Asterias* is exclusively marine and lives on most sea-coasts and in the shore waters from the tide lines to considerable depths on sand and mud. It generally prefers rocky areas where locomotion and concealment are easier than the soft, sandy or muddy bottoms. It is bottom dwelling hence called benthonic animal. They also occur clinging to piers, piling, and similar solid, submerged objects offering firm attachment.

Habits: Usually *Asterias* spend most of the time quietly clinging to some objects. They are generally solitary but under certain ecological conditions such as direct sunlight or excessive drying many individuals may aggregate at some place for protection. *Asterias* is strongly positive to light but most of the species are negative to light and prefer shaded areas. They usually remain quiet in the day time. All the starfishes are carnivorous and voracious feeders. They are inactive during the day but at night they are much more active. Inspite of having a hard integument, it can bend and twist in a variety of ways. Some species also exhibit biological relationships like parasitism and commensalism. They also show remarkable power of autotomy and regeneration.

5.4.1 External Characters

Shape and Size: The body of *Asterias* is star-shaped flattened in the oral aboral axis, and radially symmetrical, pentamerous in arrangement. It consists of a central pentagonal central disc form which radiate out five elongated, tapering, symmetrically arranged projections called rays or arms. There are usually five arms but some forms may have more e.g. the sunstar *(Solaster)* for instance has 11-13 arms. In some genera, the number of arms vary from 7 to 40. The arms are not sharply marked off from the disc, but taper towards their distal free ends. The length of the arms is two or three times the diameter of the disc.

The axes of the arms are termed radii and the regions of the central disc between the arms are the interradii. The smallest starfish shows a diameter of 1 cm whereas largest starfish is of 80 inches.

The body has an oral or actinal surface on which mouth is situated. The oral (lower) surface is normally kept towards the substratum. The upper or aboral or abactinal surface is convex covered with spines of various length. At the centre of aboral surface, a minute opening is situated called anus. This surface also bears madreporite.

Colour: *Asterias* is usually brightly coloured yellow, brown or orange but also there are many colour variations. The aboral surface is directed upwards and much darker in colour which is bright orange but varies to almost straw colour.

Oral Surface: The lower surface i.e. the one normally kept towards the substratum is flat called oral or actinal surface. The oral surface bears the structures such as mouth, ambulacral grooves, ambulacral spines, tube feet, eyes and tentacles.

(1) Mouth: The mouth or actinosome is a circular aperture situated at the centre of the oral surface of the central disc. It is surrounded by a soft membrane called peristomial membrane or peristome and guarded by five groups of oral spines or mouth papillae.

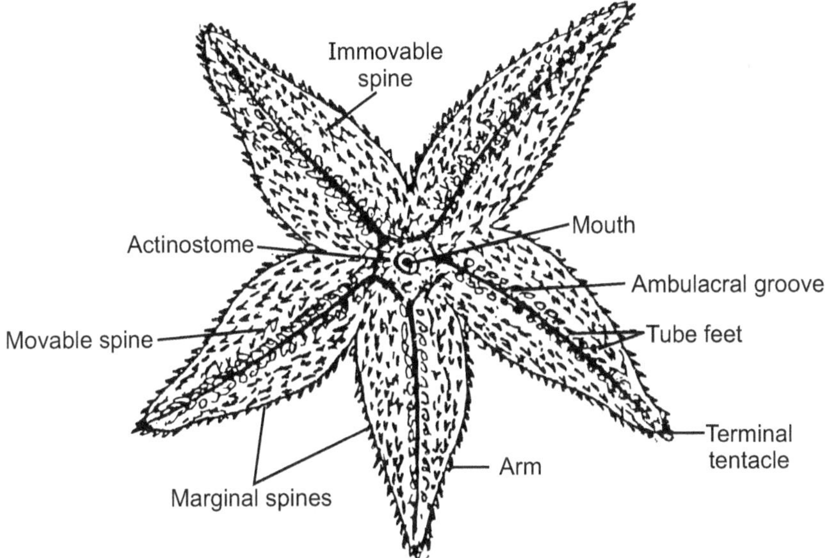

Fig. 5.7: *Asterias:* **Oral View**

(2) Ambulacral Grooves: From the five corners of mouth or actinosome radiate out five narrow grooves called ambulacral grooves. Each ambulacral groove of variable width runs along the middle of the oral surface of each arm upto its lip. Each groove shows two rows of tube feet, provided with terminal discs.

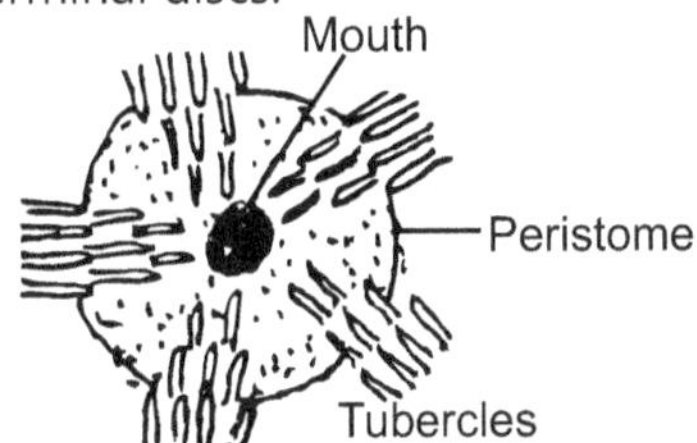

Fig. 5.8: *Asterias:* **Mouth Armature**

(3) Ambulacral Spines: Each ambulacral groove is bordered and guarded from the lateral side by 2 or 3 rows of movable calcarious ambulacral spines. These spines are capable of closing over the groove. Near the mouth, these spines become larger, stouter and are grouped together to form mouth papilla. Outside the ambulacral spines are three rows of immovable spines. Then there is a row of similar spines along the borders of the arms separating the oral from the aboral surface. The spines on the aboral surface are whitish and immovable and occur in irregular radial rows.

(4) Tube Feet or Podia: Each ambulacral groove contains two double rows of soft, thin walled, extensile, tubular structures called tube feet. The tube feet ends in a sucker and sucker works as suction cup for the firm attachment on the surface to which it is applied. Tube feet are multipurpose organs. They are mainly used for locomotion and food capturing and also help in respiration and adherence to substratum. They are sensory too.

(5) Eyes and Tentacles: At the end of each ambulacral groove, there is a small, light sensitive, bright reddish pigment spot, the eye. Each eye is made up of several ocelli. Above the eye projects a median non-retractile process called the tentacle. There are five unpaired terminal tentacles and five unpaired eye spots. The tentacles are tactile and olfactory in function.

Aboral Surface:

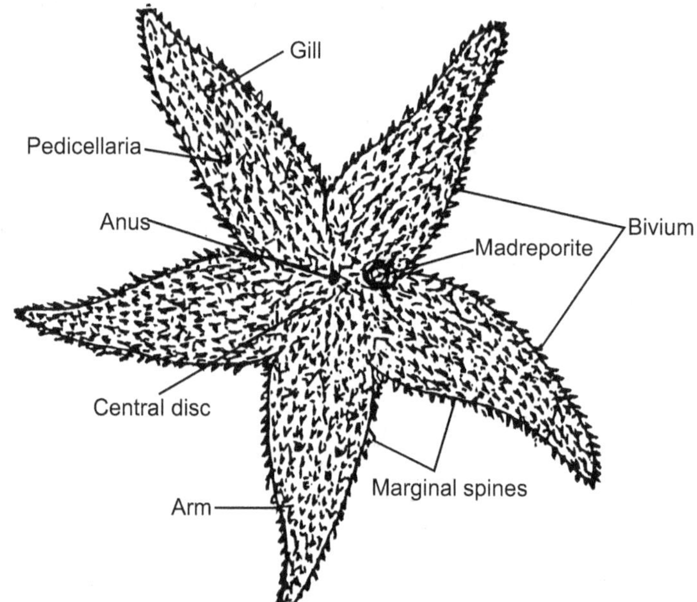

Fig. 5.9: *Asterias*: **Aboral View**

The upper convex surface is called aboral or abactinal surface which bears number of structures such as spines, dermal branchiae, anus, madreporite and pedicellariae.

(1) Spines: The entire aboral surface is also covered by short, stout, blunt and immovable calcarious spines or tubercles. They are the outgrowths of calcarious plates, the ossicles embedded in the body wall. These spines are arranged in irregular rows parallel with the axes of arms.

(2) Dermal branchiae or Papulae: Large number of minute dermal pores are present between the ossicles of integument. Though each dermal pore projects out a very small, soft, delicate, hollow, finger like membranous retractile process called dermal branchia, gill or papula. They can be completely retracted into the body. Each papula is a hollow evagi-nation of the body wall containing extension of the coelom. Since, the wall of the papula is very thin, it is presumably respiratory in function.

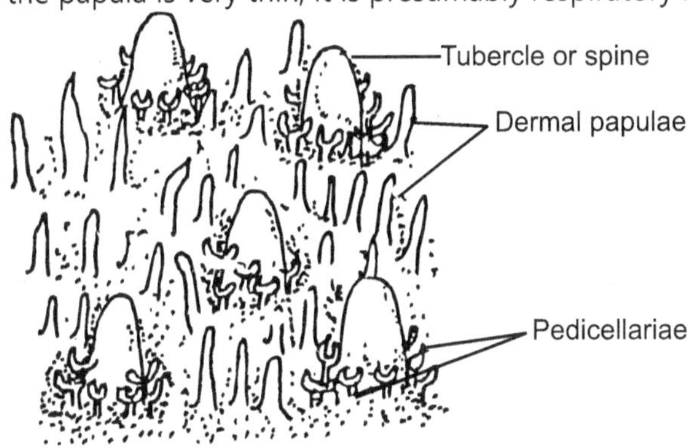

Fig. 5.10: *Asterias:* **Different Structures on the Skin**

Through its thin wall oxygen diffuses to the coelomic fluid and carbon dioxide diffuses out. The amoebocyte cells can pass out through this wall alongwith waste products collected by them in the coelomic fluid, thus performing the excretory function.

(3) Anus: It is a small aperture hardly visible to the naked eye. It lies nearly in the centre of the aboral surface, being slightly displaced towards the inter-radius next to that occupied by the madreporite. Echinoderms are the only animals that have an anus in a dorsal position.

(4) Madreporite: It is flat, nearly circular, small but conspicuous button like structure called madreporite. It is situated on the aboral surface eccentrically.

The two rays between which madreporite is placed are called the bivium and three remaining rays trivium. The madreporite is a sieve like porous plate and leads to the stone canal of the water vascular system. Usually only single madreporite is present in the starfish but sometimes more than one madreporites also occur, because of increased number of arms. The asymmetrical position of the madreporite converts the radial symmetry of the animal into a bilateral symmetry because it is only one vertical plane which passes through the middle of the madreporite as well as the middle of the opposite arm, that divides the body into two similar halves.

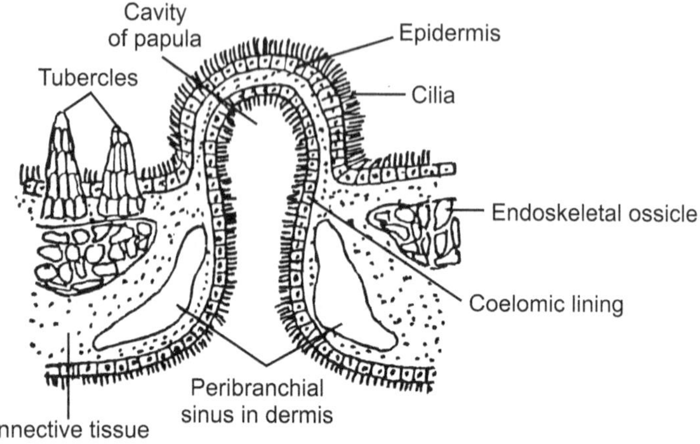

Fig. 5.11: *Asterias:* Section through Papula

5.4.2 Pedicellariae

The pedicellariae are the modified spines that occur in the space between the spines or in clumps around the bases of the spines all over the body. They are microscopic pincer-like or jaw-like bodies.

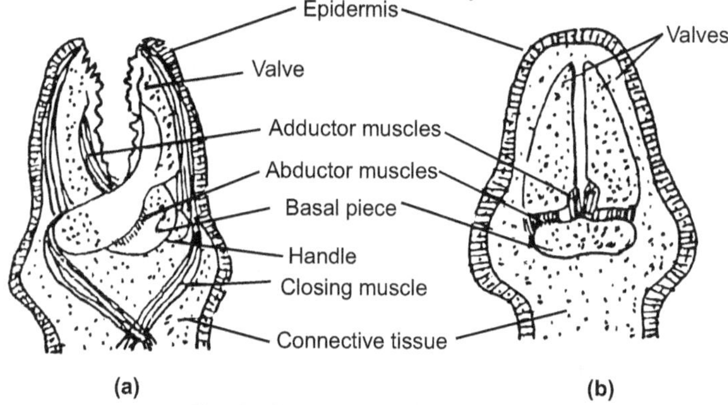

Fig. 5.12: *Asterias:* Pedicellariae

(a) Crossed or Scissors type. (b) Straight or Forceps type

Structure of Pedicellariae: The stalked or pedunculate type of pedicellariae are found in the genus *Asterias*. Each pedicellaria consists of a short, flexible and fleshy stalk, but there is no internal calcarious support. The stalk bears three calcarious plates or ossicles, a basilar plate at its top and two jaws or valves. The jaws are articulated with the basilar plate and serrated along their opposed edges. The pedicellariae having three calcarious pieces and a stalk are called forcipulate pedicellariae. They are covered with epidermis which is richly supplied with sensory and gland cells.

Types of Pedicellariae: There are two types of forcipulate pedunculate pedicellariae found in *Asterias*. These are forceps or straight type and scissors or crossed type.

(1) Forceps or Straight type: It is a simple type in which two jaws are more or less straight and attached basically to the basal piece. When pedicellariae is closed the jaws remain parallel and meet throughout their length like a forcep. The jaws can be opened or closed by muscles. The two jaws are operated by two pairs of adductor muscles to close them and one pair of abductor muscles to open them.

(2) Scissors or Crossed type: These pedicellariae are relatively small and are arranged in rings round the white spines on the aboral surface. In this type, the basal ends of the two jaws are curved and cross each other like the mandibles of a cross bill, so that the basal piece is enclosed between their crossed portions. The movement of the jaws is effected by three pairs of adductor muscles and one pair of abductor muscles. The abductor muscles originate on opposite ends of the basal ossicle and insert on the neighbouring crossed parts of the jaws. An elastic ligament is present in the stalk which bifurcates for attachment to the outer surface of the basal end of each jaw. This type of pedicellariae functions like a pair of scissors. Sessile pedicellariae also occur on the body of *Asterias*.

Functions of Pedicellariae: The pedicellariae perform different functions. They are useful for the protection of delicate skin, gills or papulae and keep the body surface free from debris and foreign organisms. They also serve as defensive and offensive organs. In some starfishes the pedicellariae are said to help in capture of small prey. They are also sensitive to contact.

5.4.3 Digestive System

The digestive system consists of an alimentary canal and digestive glands.

Alimentary Canal: The alimentary canal is complete. It is tubular and short extending vertically along the oral - aboral axis in the central disc. Due to flattening of the body it is much shortened, but wide at places and considerably modified. It comprises the mouth, oesophagus, stomach, intestine and anus.

(1) Mouth: The mouth is a small opening in the central disc on the oral surface situated in the centre of the peristomial membrane. It is surrounded by spincter muscles and radial muscle fibres and capable of great expansion and retraction.

(2) Oesophagus: The mouth leads into a short wide, vertical tube called oesophagus. It opens aborally into stomach.

(3) Stomach: The stomach is largest part of the alimentary canal. It is differentiated into two distinct parts by a horizontal constriction, the lower cardiac stomach and the upper pyloric stomach.

(i) Cardiac stomach: It is a spacious sac like structure occupying the greater part of the central disc. It is thin walled, muscular and highly folded forming a five lobed sac one opposite each arm. It can be completely everted through the mouth by pressure of coelomic fluid caused by the contraction of body wall musculature. The retraction of cardiac stomach is brought about by five pairs of retractor muscles which are originated from the lateral sides of the ambulacral ridges. In each arm the cardiac stomach is attached to the ambulacral ridge by ligaments of muscles and connective tissues called mesenteries or gastric ligaments. These mesenteries are useful to anchor the cardiac stomach in place. The cardiac stomach is glandular and secretes mucus.

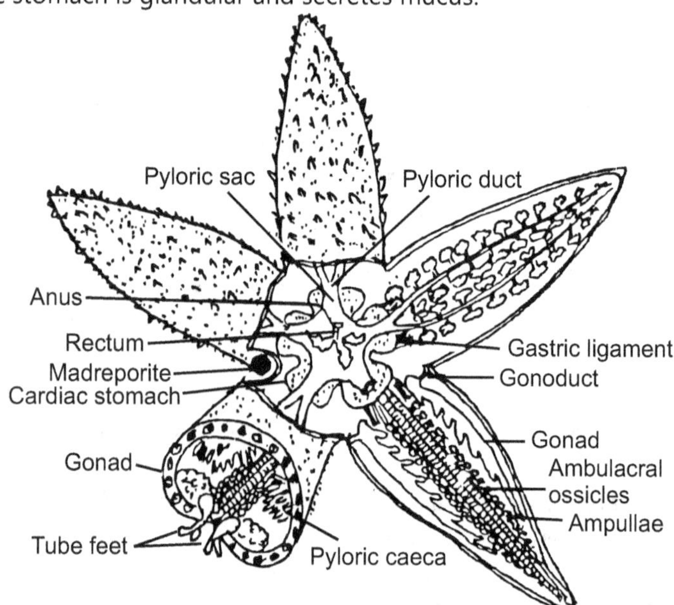

Fig. 5.13: Asterias: Alimentary canal in Aboral View

(ii) Pyloric Stomach: The pyloric stomach is a smaller, pentagonal sac in communication with cardiac stomach. Each angle of the pyloric stomach is drawn out radially into a duct which enters the corresponding arm and branches to form a pair of large appendages called as pyloric caeca or heaptic caeca or digestive glands or gastric glands. Thus, there are five pairs of pyloric caeca, one pair in each arm extending right up to its tip and each caecum is suspended from the aboral wall of the arm by two longitudinal mesenteries.

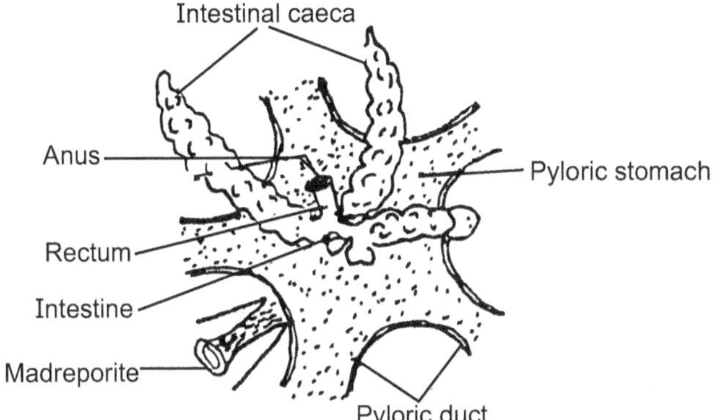

Fig. 5.14: *Asterias* : Pyloric stomach and intestine

(4) Intestine: From the pyloric stomach ascends short, narrow, five sided intestine, that runs straight upward to open at the anus. It gives off 2 or 3 little hollow diverticula called intestinal or rectal caeca before opening at the anus. The intestinal caeca are brown in colour and secrete brownish secretion. They are probably excretory in function.

(5) Anus: Intestine opens on the aboral surface by a small opening on the central disc called anus. It is situated slightly away from the centre.

Digestive Glands: The digestive glands of *Asterias* are five pairs of long, brownish or greenish bodies called the pyloric caeca. In each pyloric caecum, the hollow axis gives off laterally two series of small hollow branches, each terminating into a number of small bladder like pouches or lobules.

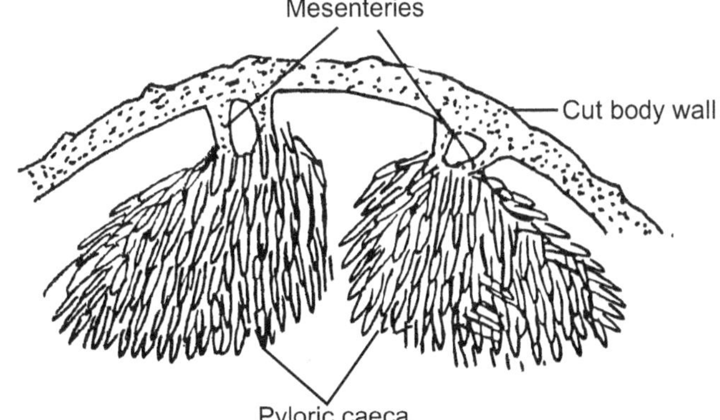

Fig. 5.15: *Asterias* : Aboral part of arm in T. S. showing mesenteries of pyloric caeca

The pyloric caeca are lined by ciliated columnar epithelium in which glandular and storage cells are also present. The glandular cells are granular and secrete a digestive juice similar to that of pancreatic juice of vertebrates. It contains enzymes like proteases, amylases and lipases that digest proteins, starches and fats respectively. There are four types of cells present in the lining of ciliated columnar epithelium of pyloric caeca. The current producer cells are with flagella and maintain steady circulation of the fluids and digested food in the cavities of caeca, the mucous cells produce mucous, the secretory or granular cells secrete digestive enzymes and the storage cells store reserve food material like lipids, glycogen and sugar protein complex. The function of pyloric caeca is like the pancreas of vertebrates.

5.4.4 Food and Feeding Mechanism

Food: Asterias is a carnivorous animal and feeds voraciously on worms, crustaceans, snails, bivalves, small starfishes, sea-urchins and small fishes. The bivalves however, form its favourite food. It also feeds on dead animals and thus, acts as a scavenger in the sea. It can survive for several months without food.

Feeding Mechanism: The mode of feeding in Asterias is of unusual type. Small sized prey is directly swallowed through the mouth into the stomach. The prey is held by the tube feet and cardiac stomach is everted and wrapped round it. Enzymes are poured out on to the prey and when digestion is complete, stomach is withdrawn in the body with the digested food.

The prey like oysters, mussels or clams are captured by *Asterias* in a very interesting manner. It creeps over the mussel and holding it with tube feet orients it to bring the free margins of the shell close to its mouth.

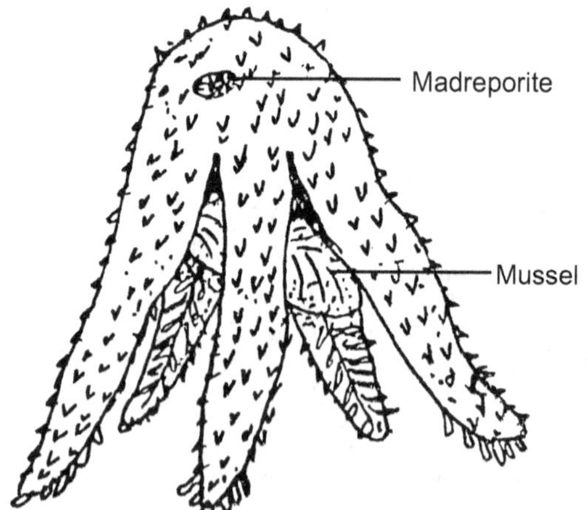

Fig. 5.16: *Asterias* : Showing feeding mechanism

Asterias arches its body over the prey, assuming a characteristic humped or umbrella like posture. The more proximal tube feet firmly grip both valves of prey while more distal ones are firmly attached to the substratum. By the contraction of body muscles the coelomic fluid exerts pressure on the stomach and cardiac stomach is everted through the mouth. The tube feet gripping the valves of mussel exert a steady pull as the muscles in the arms contract. The mussel has a very powerful set of muscles which close the shell, but the pull exerted by *Asterias* is greater than what adductors of mussel can resist. If the prey is large it may resist the pull for sometime. During this act *Asterias* use its numerous tube feet. When one arm gets tired, it is put to rest and another arm is used. As per the principle of muscle physiology, a muscle can't maintain a continuous state of contraction for a long period. Sooner or later the adductor muscle of bivalve becomes fatigued or exhausted and finally relaxes so that the shell opens. *Asterias* now inserts its already everted cardiac stomach into the mantle cavity of the mussel and pours out powerful proteolytic enzymes which dissolve the soft viscera of prey.

Physiology of Digestion: Digestion in *Asterias* is extracellular as well as intracellular. The proteolytic enzymes convert proteins into diffusible peptones, starch into maltose and fats into fatty acids and glycerol. The cilia of the cardiac and pyloric stomach set up strong currents that carry the digested food into the stomach and then pyloric caeca. When the digestion is partially completed, *Asterias* withdraws its stomach along with the digested food by means of its retractor muscles and closes its mouth by contraction of spincter muscle. It then moves away discarding empty shell. Remaining digestion of food occurs in pyloric stomach and pyloric caeca.

Digested food is absorbed through the walls of digestive tract into coelom and gets distributed to all parts of the body through coelomic fluid. The food is chiefly absorbed in the pyloric caeca, which also stores the surplus food.

Undigested matter is largely thrown out through the mouth and very little matter passes out of the anus.

5.4.5 Water Vascular (Ambulacral) System

Presence of water vascular system is a unique feature of the anatomy of all the echinoderms. It is derived from the coelom, from which it separates during development of the animal. Thus, it is the modified part of the coelom and consists of a system of canals lined with ciliated epithelium and filled with sea water, containing certain corpuscles. Its original function is concerned with obtaining food and it plays a vital role in the locomotion of the animal. The water vascular system is a sort of hydraulic-pressure mechanism and consists of madreporite, stone canal, ring canal, radial canals, Tiedeman's bodies, polian vesicles, lateral canals and tube feet.

(1) Madreporite: The water vascular system starts from the madreporite, which is hard, rounded and calcarious plate like structure present on the aboral surface of the central disc of the body. The madreporite is situated in an interradial position i.e. between the bases of two of the arms of bivium. Its circular surface is marked by numerous fine, straight or wavy radiating grooves or furrows, having at their bottom minute ciliated pores. Each furrow contains about 200 to 250 pores, which lead into very short and fine tubes called the pore canals. All these pore canals unite to form the collecting canals. The latter in turn open into a small sac like structure called, the ampulla lying just below the madreporite.

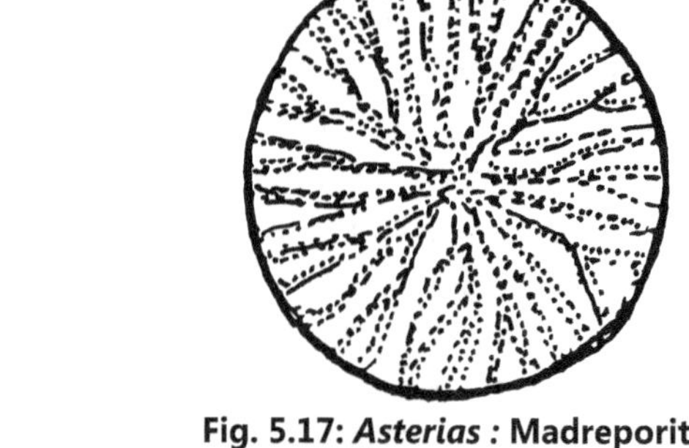

Fig. 5.17: *Asterias* **: Madreporite**

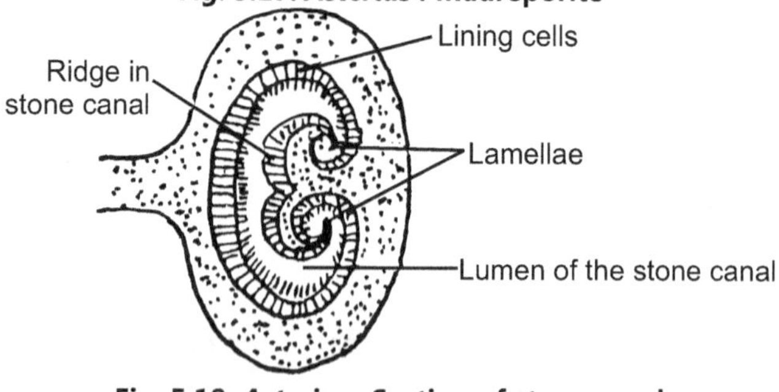

Fig. 5.18: *Asterias* **: Section of stone canal**

2.   Stone Canal: The ampulla opens into a vertical, 'S' shaped stone canal or madreporite canal which extends downwards (orally). Its wall is supported by a series of calcarious rings hence called stone canal. The lumen of the stone canal is lined by very tall flagellated or ciliated cells. The movements of the cilia draw the water current into the canal. On one side its wall projects into its cavity as a ridge that bifurcates into two lamellae rolled spirally. The ridge and its lamellae occupy a greater part of the lumen of the stone canal. The stone canal is completely surrounded with an axial organ, a wide, thin walled tubular coelomic sac, called the axial sinus, which forms a part of the perihaemal system. The stone canal at its oral end opens into a pentagonal or circular canal.

(3) **Ring Canal:** The ring canal is a wide, five sided or pentagonal, ring like canal situated around the mouth. It lies just to the inner side of the peristomal ring of ossicles and directly above the outer hyponeural ring sinus.

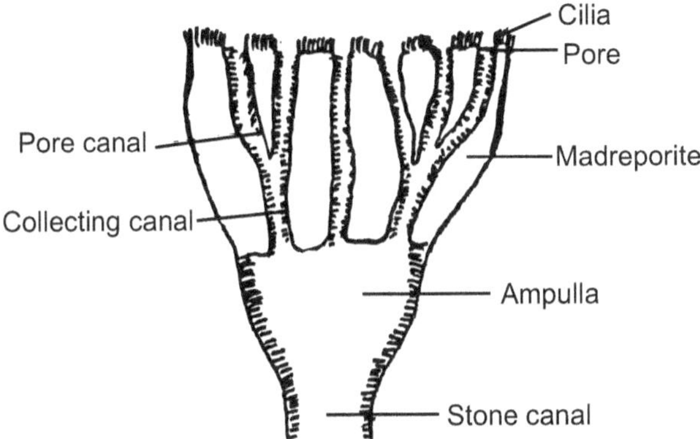

Fig. 5.19: *Asterias* **: V. S. of madreporite**

(4) **Tiedemann's bodies :** The ring canal gives off inter-radially a pair of small vesicles on its inner side called racemose glands or Tiedemann's bodies. These are small yellowish, irregular or rounded glandular bodies. They rest upon the peristomial ring of ossicles. There are only 9 Tiedemann's bodies, the 10th being absent and its position is taken by stone canal. The cavity of Tiedemann's-body is divided into number of chambers and opens into ring canal. Each body is hollow and folded and lumen contains coelomocytes and coagulated fluid. The exact function of Tiedemman's body is not known but it is believed that they are lymphatic glands and manufacture the amoebocytes of the water vascular system. It is also believed that they produce phagocytic coelomocytes but there is no experimental evidence in favour of this view.

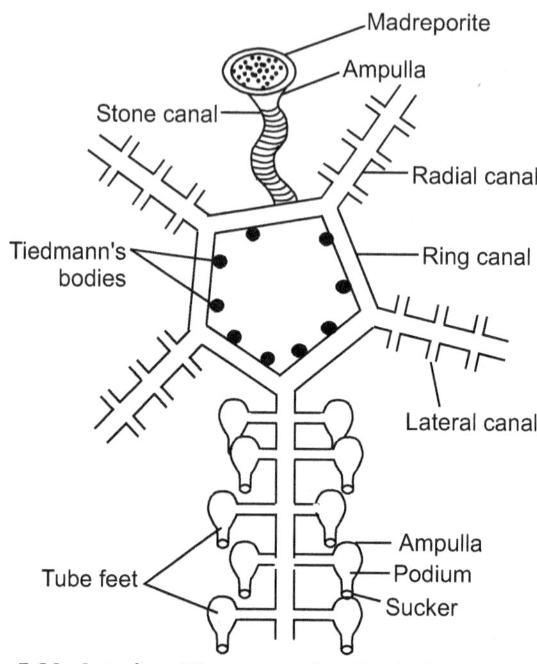

Fig. 5.20: *Asterias* : Water vascular (Ambulacral) system

(5) Polian Vesicles: In most sea stars but not in *Asterias* the ring canal gives off in each inter radius a large, thin walled, pear shaped sac called polian vesicle. Their number is variable from 2 to 4 on inter radius. They are contractile structures which store water. Polian vesicles perform different functions. They are supposed to regulate the pressure in water vascular system and manufacture amoeboid cells. Each polian vesicle is lined by a flattened epithelium, followed by a well developed circular muscle layer and stratum of connective tissue containing longitudinal muscles.

(6) Radial Canals: Ring canal from its outer surface gives off five long and ciliated radial canals into each arm. The radial canal runs upto the tip of the arm and terminates as the lumen of a small terminal median tentacle, bearing an eye at its base.

(7) Lateral Canals: During the course of radial canal it gives off on either side a series of short, narrow, transverse branches called lateral or podial canals. They are arranged alternately long and short, and open into the tube feet. The opening of the podial canal into the tube feet is provided with a valve that prevents the passage of fluid from the tube feet into the lateral canal.

(8) Tube Feet: There are actually two alternating rows of tube feet on either side in ambulacral groove of each arm. As the lateral canals along each side are alternately long and short, it gives the appearance of four rows of tube feet instead of two. Each tube feet or podium is a hollow elastic, thin walled sac like structure. The upper sac like part is called ampulla, a middle tubular podium and lower disc like structure called sucker. The ampulla, lies within the arm projecting into the coelom above the ambulacral pore. Muscle fibres present in the walls of the ampullae and the tube foot are in continuation with the lateral or podial canal. The tube feet are locomotory and respiratory organs of *Asterias*.

The various parts of the ambulacral system have a muscular wall lined internally by ciliated epithelium and covered by coelomic fluid, practically identical with the sea water. It enters through the pores of the madreporite and circulate through system. The circulation of the fluid occurs by the action of cilia that line the entire system. There is steady intake of water through the madreporite which indicates that there must be some leakage of water through tube-feet.

The water vascular system mainly helps in locomotion, adherence to the substratum and plays an important role in the respiration.

5.4.6 Autotomy and Regeneration

Asterias shows unique phenomenon called autotomy and regeneration. In nature, animal is liable to suffer an injury. If an arm is injured or held up the starfish usually casts it off near the base at fourth or fifth ambulacral ossicle. This process is called autotomy. This phenomenon is also found in many other invertebrates. The dropped arm may then be regenerated. The opening left in the side of the central disc by the broken off arm is immediately closed by the contraction of the adjacent body wall musculature for the protection of internal organs. When the process of regeneration starts, first terminal tentacle, the terminal ossicle and the eye are formed.

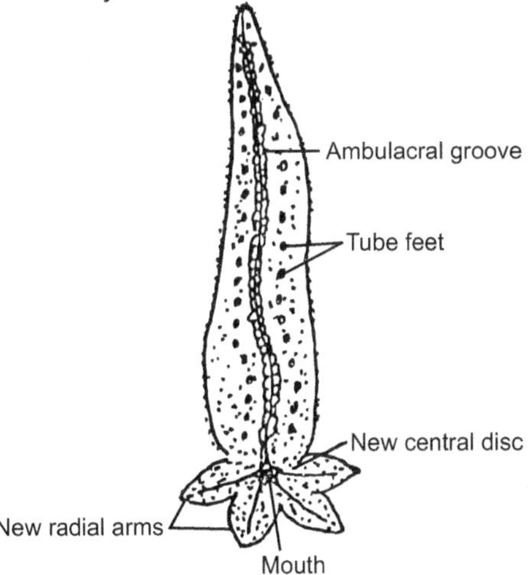

Fig. 5.21: Regeneration of Complete Animal from Cast off Arm in Linckia

The internal organs lost such as pyloric caeca, radial canal, radial nerve cold etc. grow out from their stumpy remains.

Experimentally also it is proved that even if four arms of the animal are cut off, it can survive and all four arms are regenerated in due course of time. The process or regeneration still occurs even all five arms are removed but the animal should be fed properly. In case of Asterias the isolated arm can not develop the central disc and other arms. In some species e.g. Linckia, however, a single isolated arm can grow into a complete animal. But in some species at least one fifth of disc must remain attached to an arm for regeneration. Therefore some species cast off their arms as a method of reproduction

Adaptations in *Asterias*:

Asterias shows the following adaptations:

1. It is benthonic (bottom dwelling) and nocturnal animal so it get protection from predators.
2. Body has hard integument and spines.
3. Pedicellariae useful for cleaning the body surface and also for capturing food material. They serve as organs of offense and defence.
4. *Asterias* show remarkable power of autotomy and regeneration which helps in replacing injured arms.
5. Locomotion is by tube feet so that it can explore new feeding areas.
6. Digestion occurs partly outside the body so it can feed on large sized prey.
7. It can survive for longer time without food.
8. Large number of eggs are laid by female which helps for continuation of the race.
9. Development is indirect i.e. larval stages are present useful for dispersal.

Some species e.g. Linckia, however, a single isolated arm can grow into a complete animal. But in some species at least one-fifth of disc must remain attached to an arm for regeneration. Therefore some species cast off their arms as a method of reproduction.

5.5 PEDICELLARIAE IN ECHINODERMS

Pedicellariae: The pedicellariae are the modified spines that occur in the space between the spines or in clumps around the bases of the spines all over the body. They are microscopic pincer-like or jaw-like bodies.

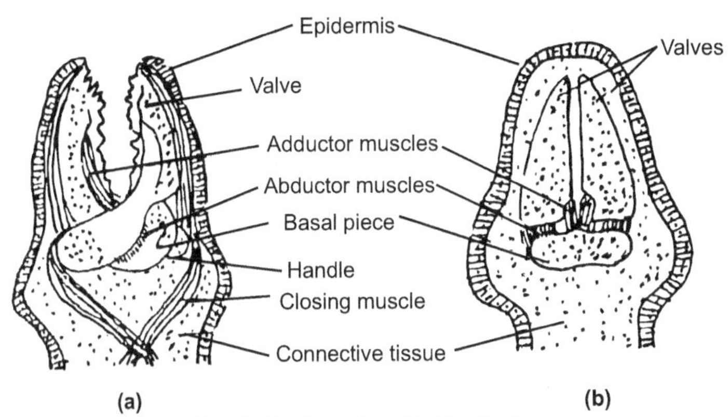

Fig. 5.22: *Asterias*: Pedicellariae
(a) Crossed or Scissors type. (b) Straight or Forceps type

Structure of Pedicellariae: The stalked or pedunculate type of pedicellariae are found in the genus *Asterias*. Each pedicellaria consists of a short, flexible and fleshy stalk, but there is no internal calcarious support. The stalk bears three calcarious plates or ossicles, a basilar plate at its top and two jaws or valves. The jaws are articulated with the basilar plate and serrated along their opposed edges. The pedicellariae having three calcarious pieces and a stalk are called forcipulate pedicellariae. They are covered with epidermis which is richly supplied with sensory and gland cells.

Types of Pedicellariae: There are two types of forcipulate pedunculate pedicellariae found in *Asterias*. These are forceps or straight type and scissors or crossed type.

(1) Forceps or Straight type: It is a simple type in which two jaws are more or less straight and attached basely to the basal piece. When pedicellariae is closed the jaws remain parallel and meet throughout their length like a forcep. The jaws can be opened or closed by muscles. The two jaws are operated by two pairs of adductor muscles to close them and one pair of abductor muscles to open them.

(2) Scissors or Crossed type: These pedicellariae are relatively small and are arranged in rings round the white spines on the aboral surface. In this type, the basal ends of the two jaws are curved and cross each other like the mandibles of a cross bill, so that the basal piece is enclosed between their crossed portions. The movement of the jaws is effected by three pairs of adductor muscles and one pair of abductor muscles. The abductor muscles originate on opposite ends of the basal ossicle and insert on the neighbouring crossed parts of the jaws. An elastic ligament is present in the stalk which bifurcates for attachment to the outer surface of the basal end of each jaw. This type of pedicellariae functions like a pair of scissors. Sessile pedicellariae also occur on the body of *Asterias*.

Pedicellariae in Echinus (Sea-Urchin):

On the body of sea urchin large number of minute pedicellariae are abundantly present. They are located between the bases of the spines and on the peristome and also upon the periproct. The peristome is thin, soft, flexible and circular membrane which surrounds the mouth, whereas the periroct is circular and leathery membrane which covers the aboral pole. There are 4 different kinds of pedicellariae occurring in sea urchins.

Each pedicellaria is made up of three parts, namely head, jaws and stalk. Head is composed of generally three (sometimes 2, or 5) movable and opposing jaws or blades. These blades or jaws are mounted on a stalk. The stalk is joined to the body by a ball and socket joint. The jaws are somewhat strong and stout because each jaw is supported by an internal calcarious piece which gives particular shape to the jaw. The stalk which possess the jaws is also internally supported by skeletal rod. Hence pedicellariae remains straight on the body.

The jaws articulate against each other and their closing and opening is operated by the muscles.

In sea urchin following four types of pedicellariae are found.

(1) Gemmiform or globiferous pedicellaria: These pedicellaria have very stiff stalks and round or globular heads. Each jaw has poison gland. Each jaw is strong and curved with terminal teeth. There are abductor and flexor muscles useful for closing and opening of the jaws. In between the jaws there is valve or endoskeletal piece present. These pedicellaria are serving as weapons of defence against large enemies. These are also found in *Strongylocentrous*.

(2) Tridentate or tridactyle pedicellariae: These are most common and largest pedicellaria found in sea-urchin. They have flexible stalk and elongated serrated tapering jaws. The jaws are meeting at the distal end. Two lateral and one central long jaws form trident structure. The inner margin of each jaw is saw blade like for cutting the body of enemies into small pieces. The stalk is not very long but short and internally supported by skeletal rod. The stalk is narrow at the both ends and broader in the middle.

The jaws are also equipped with muscles for its operations. These pedicellaria can crush the small foes to death.

(3) Ophiocephalous pedicellaria: These are smaller pedicellaria with broad and flexible stalk. The jaws are short and broad with teeth and blunt tips. They are mainly found on peristome region. They are useful for capturing small animals as food. The stalk is supported internally with axial rod.

(4) Trifoliate or triphyllous pedicellaria: These are the smallest pedicellaria found in sea urchin. Their stalks are thin and flexible. The jaws are short and broad which do not meet distally. The jaws are also not pointed and serrated but petal like. They pick up and remove debris and intruders and keep the body clean. Hence these are not used as an organ of defence and offence. These are 3 jaws hence they are called trifoliate pedicellariae.

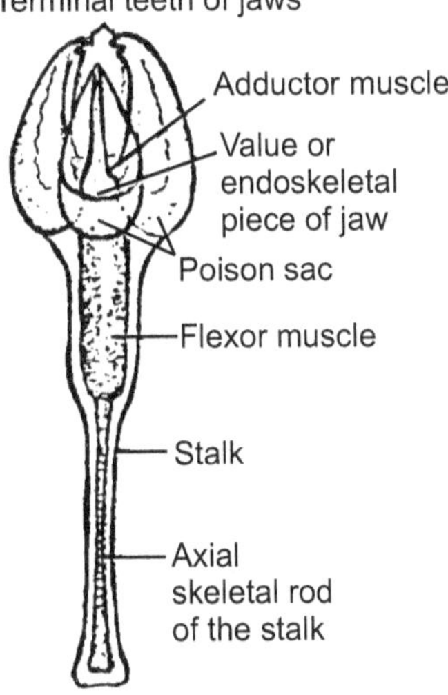

(A) Gemmiform pedicellaria of *Strongylocentrous*

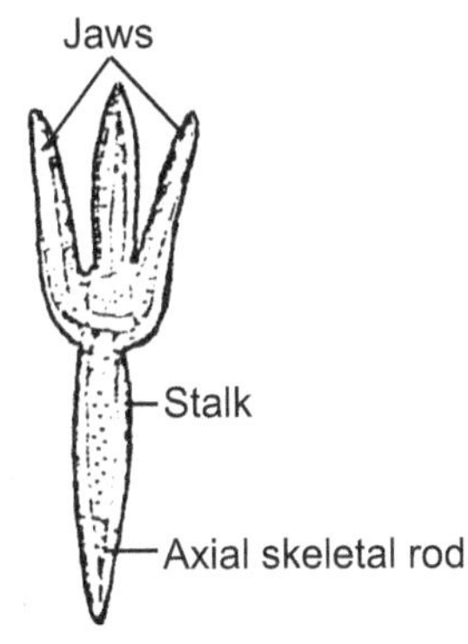

(B) Tridictyle pedicellaria of *Echinus acutus*

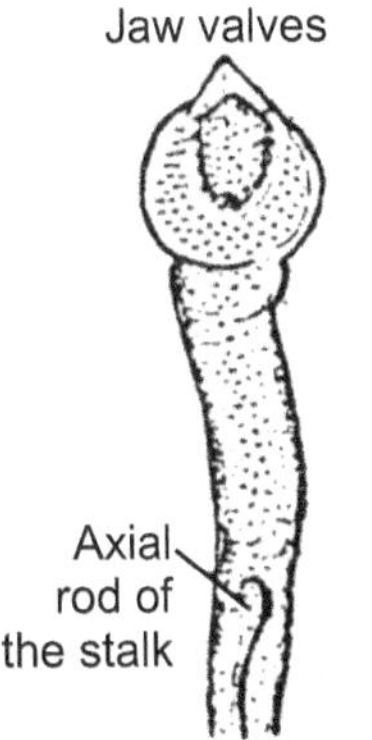

(C) Ophiocephalous pedicellaria of *Echinus acutus*

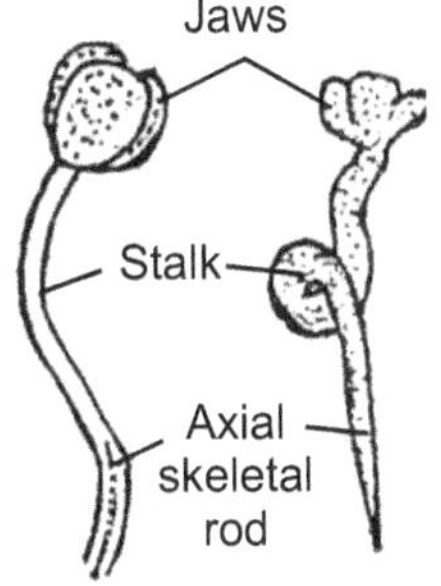

(D) Trifoliate pedicellariae
A – Of *Tripneustes*
B – Of *Echinus*

Fig. 5.23 : Pedicellariae in sea-urchin

Functions of Pedicellariae: The pedicellariae perform different functions. They are useful for the protection of delicate skin, gills or papulae and keep the body surface free from debris and foreign organisms. They also serve as defensive and offensive organs. In some starfishes the pedicellariae are said to help in capture of small prey. They are also sensitive to contact.

5.6 ECONOMIC IMPORTANCE OF ECHINODERMATA

The economic importance of the echinoderms in their relation to the lower animal is uncertain. But they are of considerable importance to man. The starfishes and sea cucumbers are of much economic value to man.

1.　As a fertilizer: The dried skeletons of echinoderms have been crushed and used as fertilizer for crops, as they are quite rich in calcium and nitrogenous content.

2.　As a food: The dried powder of skeleton of echinoderms is used as a food and lime supply for poultry. Sea cucumbers are commonly used as food by man in Pacific Islands and the shores of south Pacific oceans and China. These animals are boiled and dried in sun and sold in market which is used for soup. In the orient it is a culinary delicacy and imparts distinct flavour when cooked in certain dishes. In America canned minced sea cucumber has been used, especially in soup, for several years. The ovaries of sea urchin and the eggs of starfish are eaten either raw or cooked, by the people of Europe, tropical regions of South America.

3.　As a scavengers: Starfishes, sea cucumber and sea urchins are good scavengers. They ingest mud and sediment of sea bottom and extract the contained organic food. Thus they help in decomposition processes of organic material continuously deposited at the bottom. They feed on sea weeds, small crustaceans, molluscs, tube worms, dead animal matter and bottom debris.

4.　Harmful echinoderms: They are predator animals and destroy dams, oysters and other marine molluscs which serve as human food. Starfishes are able to open shells of bivalves, oysters and kill them in large number. They cause severe damage to the commercial oyster beds. Therefore, owners of oyster beds use various techniques to capture the starfishes. They are destroyed in hot water carried ashore to dry up. They are also killed by sprinkling quick lime on oyster beds which is not harmful to oysters. Starfishes control the non-commercial molluscs.

POINTS TO REMEMBER

- Echinoderms are exclusively marine.
- They are gregarious, mostly free-living, creeping and some pelagic.
- Water vascular system is present.
- No sexual dimorphism.
- External fertilization.
- Echinoderms are classified in Asteroidea, Holothuria, Crinoidea, Echinoidea.
- Water vascular is unique feature of the echinoderms.
- Main function of this system is for obtaining food and locomotion.
- Madreporite, Stone canal, Ring canal, Tiedemann's bodies, Polian vesicles, Radial canals, Lateral canals, Tube feet are the important organs of the water vascular system.
- *Asterias* shows oral and aboral surface.
- Tube feet are locomotory organs.
- Dermal branchial are present on the body for respiration.
- On the body *Asterias* pedicellariae are present for the protection.
- Pedicellariae are of different types.
- *Asterias* has digestive system with different parts.
- Water vascular system is useful for obtaining food and locomotion.
- Echinoderms are economically important animals.

EXERCISE

1. Give an account of general characters of Phylum-Echinodermata. Give an outline of classification of Phylum-Echinodermata.
2. Give distinguishing characters of the following classes with suitable two examples.
   - (a) Asteroidea
   - (b) Ophiuroidea
   - (c) Echinoidea
   - (d) Holothuroidea
   - (e) Crinoidea
3. Write short notes on :
   - (a) Tubefeet
   - (b) Madreporite
   - (c) Water vascular system
   - (d) Stone canal
   - (e) Radial canal
   - (f) Water vascular system
   - (g) Autotomy and regeneration
   - (h) External characters of *Asterias*
   - (i) Types of Pedicellaria
4. Give an account of water vascular system in *Asterias*.
5. Describe the various types of Pedicellaria in Echinoderms and add a note on their functions.
6. Give an account of economic importance of Echinoderms.
7. Describe the digestive system of *Asterias*.
8. Give an account of external characters of *Asterias*.

MODEL QUESTION PAPER

Marks : 35   **Duration : 2 Hours**

Note :

    (a) Question No. 1 is compulsory.

    (b) Attempt any three questions from Q. 2 to Q. 5.

    (c) Questions 2 to 5 carry equal marks.

---

1. Attempt any *five* of the following :   **(5)**

    (a) Which disease is caused by *Ascaris and Wuchereria* ?

    (b) 'Earthworm is friend of farmers'. Explain.

    (c) 'All annelids are true coelomate animals'. Explain.

    (d) Why ecdysis or moulting occurs in Arthropods ?

    (e) Which insect produces the lac ?

    (f) Which are the sense organs of Molluscs ?

2. (A) What are the Aschelminthes ? Give an account of salient features of phylum Aschelminthes.   **(6)**

    (B) Describe in brief economic importance of Annelida.   **(4)**

3. (A) What is Arthropoda ? Describe different salient features of Phylum Arthropoda.   **(6)**

OR

    Give an account of classification of Phylum Annelida upto classes with example.

    (B) Give an account of economic importance of Lac insect.   **(4)**

4. (A) Give an account of water vascular system of *Asterias*.

OR

    Describe digestive system of *Asterias*.   **(6)**

    (B) Describe the economic importance of Mollusca.   **(4)**

5. Write short notes on any *four* of the following :   **(10)**

    (A) Economic importance of Nematoda.

    (B) Vermicomposting

    (C) Mouth parts of female anopheles mosquito

    (D) Useful insect honey bee

    (E) Economic importance of Echinodermata

    (F) Pedicillariae and their functions.

✳✳✳